# Biotechnology

## for Sustainable Agriculture

# Biotechnology
## for Sustainable Agriculture

*Editors*

**S. Venku Reddy**

**M. Suryamani**

**K.L. Srivastava**

**T. Goverdhan Reddy**

BSP **BS Publications**
A unit of **BSP Books Pvt., Ltd.**

4-4-309/316, Giriraj Lane, Sultan Bazar,
Hyderabad - 500 095
Phone : 040 - 23445605, 23445688

**BSP BS Publications**
A unit of **BSP Books Pvt., Ltd.**

4-4-309/316, Giriraj Lane, Sultan Bazar,
Hyderabad - 500 095
Phone : 040 - 23445605, 23445688
**e-mail : info@bspbooks.net**

**ISBN 978-93-52300-05-1 (HB)**

# Foreword

The elucidation of the double-helix structure of the De-oxy ribose Nucleic Acid (DNA) molecule in 1953 by Drs. James Watson, Francis Crick, Maurice William and Franklin Rosalind marked the beginning of what is now known as *the new genetics*. Research during the last 51 years in the fields of molecular genetics and recombinant DNA technology has opened up new opportunities in agriculture, medicine, industry and environment protection. The ability to move genes across sexual barriers has led to heightened interest in the conservation and sustainable and equitable use of biodiversity, since biodiversity is the feedstock for plant, animal and microbial breeding enterprises.

Considerable advances have been made during the last 25 years in taking advantage of the new genetics in the areas of medical research, production of vaccines, sero-diagnostics and pharmaceuticals for human and farm animal health care. The production of novel bioremediation agents as for example, the production of a new Pseudomonas strain for clearing oil spills in oceans, rivers and lakes by Dr. Anand Chakraborty, is also receiving increasing attention.

There has also been substantial progress in agriculture, particularly in the area of crop improvement through the use of molecular marker assisted breeding, functional genomics, and recombinant DNA technology. A wide range of crop varieties containing novel genetic combinations are now being cultivated in USA, Canada, China, Argentina and several other countries. A strain of cotton containing the *Bacillus thuringiensis* gene (Bt Cotton), which has resistance to boll worms, is now under cultivation in India based on both official and unofficial (illegal) releases.

There is little doubt that the new genetics has opened up uncommon opportunities for enhancing the productivity, profitability, sustainability and stability of major cropping systems. It has also created scope for breeding crop varieties tolerant/resistant to biotic and abiotic stresses. It has led to the possibility of undertaking anticipatory breeding to meet potential changes in temperature, precipitation and sea level as a result of global warming.

While the benefits are clear, there are also many risks when we enter the territory of the unknown and unexplored. Such risks relate to potential harm to the environment and to human and animal health. There are also equity and ownership issues in relation to biotechnological processes and products. The following issues are the major areas of concern to the public and policy maker.

(a) *What is inherently wrong with the technology ?*

Is the science itself safe, as for example, the use of selectable marker genes conferring antibiotic or herbicide resistance?

(b) *Who controls the technology ?*

Will it be largely in the private sector? If the technology is largely in the hands of the private sector, the overriding motive behind the choice of research problems will be private profit and not necessarily public good. If this happens, "orphans will remain orphans" with reference to choice of research priorities. Crops being cultivated in rainfed, marginal and fragile environments, which are crying for scientific attention, may continue to remain neglected.

(c) *Who will have access to the products of biotechnology ?*

If the products arising from recombinant DNA technology are all covered by intellectual property rights (IPR), then the technology will result in social exclusion and will lead to a further enlargement of the rich-poor divide in villages.

(d) *What are the major biosafety issues ?*

There are serious concerns about the short and long term impact of GMOs on the environment, biodiversity and human and animal health.

Thus, there is need for transparent and truthful risk-benefit analysis in relation to GMOs, on a case-by-case basis. In the coming decades, Indian farm women and men will have to produce more food and other agricultural commodities to meet home needs and to take advantage of export opportunities, under conditions of diminishing per capita availability of arable land and irrigation water and expanding abiotic and biotic stresses. The enlargement of the gene pool with which breeders work will be necessary to meet these challenges. Recombinant DNA technology provides breeders with a powerful tool for enlarging the genetic base of crop varieties and to pyramid genes for a wide range of economically important traits. The safe and responsible use of biotechnology will help us to meet the challenges ahead, including those caused by climate change. At the international level, the Cartagena Protoeol on Biosafety provides a framework for risk assessment and aversion. At the national level, there is need for a regulatory mechanism, which inspires public, political and professional confidence.

The present book titled "Biotechnology for Sustainable Agriculture" is a timely contribution. We are indebted to Prof S Venku Reddy and to his distinguished colleagues for this labour of love.

**Prof. M.S. Swaminathan**

*Chairman, M. S. Swaminathan Research Foundation,*
*Chennai (Madras), India.*
*May, 2004*

# Preface

There is vast potential for biotechnological applications in the field of Agriculture. Although there is revolutionary progress in the scientific aspects of biotechnology during the recent years, the application of this technology for solving the problems of farming community in India, particularly the resource poor farmers, is quite low. Improving public awareness on issues surrounding application of biotechnology for developing sustainable agriculture in India is therefore an extremely important task. In this context, PRDIS and BTU jointly organised a series of workshops in 2001-02 for various stakeholders to discuss on issues related to biotechnology for sustainable agriculture and to suggest strategies for further work. This book includes selected papers presented during these workshops and salient recommendations.

This publication is based on the work done under the project on "Creating Public Awareness on and Popularization of Biotechnology for Sustainable Agriculture Development" which was implemented by PRDIS with the support from the APNL Biotechnology Programme.

**The research for this publication was conducted as part of the programme 'Biotechnology for Dryland Agriculture in Andhra Pradesh' with financial support of the Research and Developing Countries Division, Ministry of Foreign Affairs, the Government of Netherlands. Responsibility for the contents and for the opinions expressed rests solely with the authors; publication does not constitute an endorsement by PRDIS or BTU or the funding agency.**

We consider it our privilege to thank Dr. M.S. Swaminathan, father of Green Revolution in India, who has graciously contributed the foreword of this book.

We express our gratitude to Dr. M.V. Rao, Chairman of APNLBP for his guidance in conducting the workshops. We wish to thank Dr. G. Pakki Reddy and Mr. V.A. Raju of BTU for their contribution in planning and conducting the workshops.

We extend our appreciation and thanks to the eminent authors who have contributed their papers for this publication.

We are grateful to Executive Committee members of PRDIS for their cooperation and encouragement.

Finally, we thank all those who have assisted in preparation of this book. We specially thank Mr. G. Amruth Rao, Dr. I.V. Subba Rao and Ms. A. Maria Victoria of PRDIS for their assistance. We hope that this book will promote informed action for utilising Biotechnology as a tool for sustainable agriculture.

***Editors***

# CONTENTS

## Unit II

## Different forms of Biotechnology



## Unit - III

## Issues in Biotechnology Applications

## Unit - IV

## Partnership and Communication Strategies

## *List of Tables*

## *List of Figures*

## Abbreviations and Acronyms

| | | |
|---|---|---|
| ANGRAU | : | Acharya N.G. Ranga Agricultural University |
| APNLBP | : | A.P. Netherlands Biotechnology Programme |
| AV | : | Audio Visual |
| ATMA | : | Agricultural Technology Management Assoication |
| BTU | : | Biotechnology Unit |
| CIMMYT | : | International Maize and Wheat Improvement Centre |
| CRIDA | : | Central Research Institute of Dryland Agriculture |
| CSIR | : | Council of Scientific and Industrial Research |
| CGIAR | : | Consultative Group on International Agricultural Research |
| CCMB | : | Centre for Cellular and Molecular Biology |
| CPMB | : | Centre for Plant Molecular Biology |
| CPRI | : | Central Potato Research Institute |
| CTRI | : | Central Tobacco Research Institute |
| DAATT | : | District Agricultural Advisory and Transfer of Technology |
| DBT | : | Department of Biotechnology |
| DOR | : | Directorate of Oilseeds Research |

| | | |
|---|---|---|
| DRR | : | Directorate of Rice Research |
| DNA | : | Deoxyribo Nucleic Acid |
| EPA | : | Environmental Protection Act |
| EU | : | European Union |
| EEI | : | Extension Education Institute |
| FAO | : | Food and Agricultural Organization of the United Nations |
| FO | : | Farmers Organization |
| GATT | : | General Agreement on Trade and Tariff |
| GDP | : | Gross Domestic Product |
| GEAC | : | Genetic Engineering Approval Committee |
| GMO | : | Genetically Modified Organism |
| GOI | : | Government of India |
| GM | : | Genetically Modified |
| HYV | : | High Yielding Varieties |
| IRRI | : | International Rice Research Institute |
| ICARDA | : | International Centre for Agricultural Research in Dry Areas |
| IARI | : | Indian Agricultural Research Institute |
| IBU | : | Interactive Bottom-up Approach |
| IBSC | : | Institutional Biosafety Committee |
| ICGEB | : | International Centre for Genetic Engineering and Biotechnology |
| ICRISAT | : | International Crops Research Institute for the Semi-Arid Topics |

| | | |
|---|---|---|
| ICAR | : | Indian Council of Agricultural Research |
| IICT | : | Indian Institute of Chemical Technology |
| IPR | : | Intellectual Property Rights |
| IPE | : | Institute of Public Enterprise |
| IPM | : | Integrated Pest Management |
| ICMR | : | Indian Council of Medical Research |
| KVK | : | Krishi Visgan Kendra |
| MANAGE | : | National Institute of Agriculture Extension Management |
| NKSE | : | Neem Seed Kernel Extraction |
| NIRD | : | National Institute of Rural Development |
| NPV | : | Nuclear Polyhedrosis Virus |
| NGDP | : | National Gross Domestic Product |
| NGO | : | Non Government Organization |
| PCR | : | Polymerase Chain Reaction |
| PRDIS | : | Participatory Rural Development Initiatives Society |
| RFLP | : | Restriction Fragment Length Polymorphism |
| TERI | : | Tata Energy Research Institute |
| TRIPS | : | Trade Related Intellectual Property Rights |
| UPOV | : | International Union for the Protection of New Varieties of Plants |
| WTO | : | World Trade Organization |
| WHO | : | World Health Organization |

# UNIT - I

## Importance of Biotechnology for Sustainable Agriculture

# 1

# Importance of Biotechnology for Sustainable Agriculture

S. V. Reddy[1], M. Suryamani[2], K. L. Srivastava[3], T. G. Reddy[4], and G. Anuradha[5]

Food grain production in India has shown remarkable increase during 1960-70s due to high input oriented "Green Revolution". This enabled the country to march from deficiency and dependency on imported food to self-sufficiency. Thanks to the Scientists, policy makers, extension workers and the farmers behind this great achievement. However, productivity increases for last few years have slowed down and there is an apprehension that future increases in food production will fall behind population growth, if the present conventional breeding technology alone is relied upon. It has become apparent world wide that intensive agricultural systems involving high inputs are not only too costly but are contributing to soil degradation, pest resistance and loss of plant diversity.

[1]Executive Director, [2]Consultant (Agricultural Extension), [3]Consultant (Natural Resource Management) [4]President, and [5]Consultant (Biotechnology), PRDIS, Hyderabad.

The promotion of sustainable agricultural systems is a crucial factor since already there are signs of:

- Degradation of fragile environment
- Demand for meeting the food needs of growing population
- An ever-growing migration
- Globally competitive economic systems
- Week management systems with low human capital

Biotechnology can play a pivotal role in development of sustainable agriculture. Biotechnology in conjunction with conventional breeding can help in increasing yields dramatically through new plant varieties resistant to pests and environmental stress, in lowering the cost of inputs, in improving nutritional value of food crops and in producing environmentally friendly pest and disease management programs. The recent scientific reports have shown that biotechnologists are capable of developing a range of crop and animal-based technologies for solving many of our agricultural programmes.

The development in molecular biology and gene technology in the past several years have produced a powerful set of tools for agriculture. The pest-resistant Bt cotton technology has already reached the farmers fields and many other technologies are in pipeline. Biotechnology has great potential to improve the productivity of animals through embryo transfer technology, genetic improvement, vaccines and disease diagnostics. IVF buffalo calf was born through embryo transfer technology. India ranks first in marine resources and aquaculture. Biotechnology played major role in improving per hectare production of carp and tiger prawns and in providing an immune stimulant against the white spot disease in prawns. The lower and intermediate forms of biotechnology like biofertilisers, biopesticides, plant tissue culture, animals vaccines also have considerable potential. They are very timely innovations to help us meet challenges of doubling our food production by the year 2025 for feeding the growing population and meeting other needs while taking greater care of the natural resources involved in food production. The Biotechnology thus can offer opportunities for increase in productivity, conservation of natural resource (biodiversity) and alleviation of poverty.

Countries like USA, Canada, Argentina, China, Australia, Mexico, Spain and South Africa have made considerable progress in application of Biotechnology in agriculture. The estimated benefit from transgenic crops globaly was reported to be nearly $ 4 billion in 2002. This is expected to increase dramatically to $ 5 billions or more by 2005.

## Research on Biotechnology

In India, research on agricultural biotechnology is carried out in IARI, TERI, ICGEB, Delhi University, Jawaharlal Nehru University, New Delhi; Bose Institute, Calcutta; ICRISAT, Hyderabad; DOR, DRR and NRCS, Hyderabad; CPRI, Shimla; CTRI, Rajamundry; IIHR, Bangalore; Mahyco Foundation, Bangalore; CPMB, Osmania University, University of Hyderabad, State Agricultural Universities, and many private firms located in Hyderabad and other parts of the country. These firms are working mostly on genetic transformation of crops and on many lower and intermediary forms of biotechnology such as Biofertilizers, Biopesticides, tissue culture and vaccines. The status of different forms of biotechnologies is given in Table 1.1.

**Table 1.1 Status of different forms of agricultural biotechnologies**

**1. Lower Forms**

| A. | **Crop** | **Pest** | **Bio-control agents** | **Status** |
|---|---|---|---|---|
| | Rice | Stem borer | *T. Japonicum* | Technology in use |
| | Maize | Stem borer | *T. Chilonis* | Technology in use |
| | Cotton | Boll worms | *T. Species* | Technology in use |
| | Ground nut | Aphids | *Cheilomenus Sex macunata* | Technology in use |
| | Red Gram | Fruit borers | *Ha NPV* | Technology in use |
| | Tobacco | Army Worms | *Sl NPV* | Technology in use |
| | Potato | Black scurf | *Verticillum Bigulattum* | Technology in use |
| | Tomato | Damping off | *Trichoderma* | Technology in use |
| **B.** | **Botanicals Pesticides / crop** | **Pest / diseases** | **Chemical** | **Status** |
| | Neem based pesticides | Sheath Blight BPH, Leaf folders ear head bug | *Azardirachta* | Technology in use |
| | Custard apple | BPH, Leaf folders ear head bug etc. | *Aceto-genens* | Technology in laboratory |
| | Pongamia | Stem rot and root rot of groundnut | *Karvengina* | Technology in laboratory |
| | Marigold | Brown rust of rice and wheat | | Technology in laboratory |

**Table 1.1** ***Contd.....***

| | Biofertilizers | Replacement of inorganic nutrients | Over all advantages |
|---|---|---|---|
| | Vermi culture 2.5 t/ha | 25-30% of the recommended fertilizers can be supplied | Physical status of the soil will be improved and soil pollution can be reduced. |
| | Azospirillum | 25-30 Kg N/Ha | Environment friendly |
| | Phosphobacteria | 30-40% of phosphorus fixed in the soil will be made available to the plant | Loss of phosphours through fixation can be over come. |

**2. Intermediate Forms**

| | | |
|---|---|---|
| (a) | Tissue culture micropropagation Teak, paulonia, Sandal wood, Neem, Banana, Coffee etc. | Elite clones developed through tissue culture are available on large scale for planting in farmers fields. Advantages are fast growth, early bearing in fruit crops etc. (Indo-American Hybrid seeds and all other private Biotech companies) |
| (b) | Vaccines :<br>Blue tongue (type-2) vaccine for sheep; EDS vaccines for poultry | |

**3. Higher Forms – Major developments in transgenic research**

**(a) Public Sector**

| Plants / crops | Transgenic(s) inserted | Purpose |
|---|---|---|
| Chickpea | Bean *alpha AI* | To generate plants, resistant to bruchids |
| Rice | *S-adenosylmethionime decarboxylase* | To generate plants, tolerant to stress |
| Cotton | *Bt. Cry gene(s)* | To generate plants, resistant to lepidopteran pests |

**Table 1.1** *Contd.....*

| | | |
|---|---|---|
| Potato | *Bt. Cry1A(b)* | To generate plants, resistant to lepidopteran pests |
| Tobacco | *Bt. Cry1A(b)* and *cry 1C* | To generate plants, resistant to Helio-coverpa armigera and spodotera litura |
| Rice | *Bar* | To generate herbicide-tolerant plants |
| Rice | *Bt. Cry1A(b)* *Xa21* | To develop plants, resistant to lepio-pteran pests, bacterial blight/disease |
| Mustard / Rapeseed | *Bar, barnase, barstar* | To generate herbicide-tolerant plants male-sterile and restorer lines for hybrid seed production |
| Rice | *Pyruvate decarboxylase* and *alcohol dehydrogenase* | To generate plants, tolerant to flooding |
| Tomato | *Ctx-B* and Tep antigens of *Vibrio cholera* | Edible vaccine development |
| Brinjal | Chitinase, glucanase and thaumatin encoding genes | To generate plants, resistant to diseases |
| Wheat | *Bar, HVAI, PIN2* | Resistance against biotic and abiotic stresses |
| Rice Pusa Basmati | *CodA, cor47* | Resistance against biotic and abiotic stresses |
| Rice | *Xa-21, cry1A (b)* | To generate plants, resistant to lepidopteran pests and bacterial and fungal diseases. |
| Brinjal | *Bt. Cry1A(b)* | To generate plant, resistant to lepidopteran pests |
| Tomato | *Bt. Cry1A(b)* | To genrate plants, resistant to lepidopteran pests |
| Cauliflower | *Bt. Cry1A(b)* | To generate plant, resistant to Plutella scylostella |

**Table 1.1** ***Contd.....***

| | | |
|---|---|---|
| Cabbage | *Bt. Cry1A(b)* | To generate plants, resistant to P.scylostella |
| Rice | *Bt. Cry1A(b), chitinase* | To generate stress-tolerant plants |
| Mustard / rapeseed | *Arabidopsis annexin gene* | To generate stress-tolerant plants |
| Mustard / rapeseed | *Choline dehydrogenase* | To generate abiotic stress-tolerant plants. |
| Potato | ACC synthase | To control fruit ripening |
| Tomato | ACC synthase | To control fruit ripening |
| Banana | ACC synthase | To control fruit ripening |
| Tobacco | Chitinase, glucanase and RIP | To generate plants, resistant to fungal attack |
| *Brassica* | Chitinase, glucanase and RIP | To generate plant, resistant to fungal attack |
| Pigeonpea | Protease inhibitor and lectin genes | To generate plants, resistant to bollworms and aphids |
| Rice | *Bt. Cry1A (b)* | To generate plants, resistant to yellow stem borer |
| Tobacco | *Bt. Crylla5* | To generate plants, resistant to Spodoptera litura |
| Rice | *Gm2* | To generate plants, resistant to gall midge |
| Chickpea | *PGIP* | To generate plants, resistant to fungal pathogens |
| Muskmelon | *Rabies glycoprotein gene* | To develop edible vaccines |
| Tomato | *Leaf curl virus* sequence | To generate plants, resistant to fungal disease |
| Tomato | Chitinase and glucanase | To generate plants, resistant to fungal diseases |
| Tomato | Coat protein gene of citrus triesteza virus | To generate transgenic citrus plants, resistant to citrus triesteza |

**Table 1.1** ***Contd.....***

| | | |
|---|---|---|
| Potato | *Ama-1** | To generate nutritionally enriched plants |
| Tomato | *OXDC** | To generate plant, resistant to fungal infection |
| Blackgram | Coat protein and replicase genes of *vigna mungo* yellow mosaic virus | To develop viral-resistant plants. |
| Blackgram | *Dianthin* and *barnase* gene or *bar* | Development of insect resistant and herbicide–tolerant plant |
| Rice | *Chitinase,β -1, 3 – glucanase* and *osmotin genes* | To develop plants, resistant to fungal infection |
| Coffee | *Chitinase, β -1, 3 glucanase* and *osmotin genes* | To develop plants, resistant to fungal infection |
| Rice | *Cry1A(b)* gene | To generate plants, resistant to lepidopteran pets, bacterial and fungal diseases. |
| Cotton | *Cry1E* and *Cry1C* with terminal altered at C-end | To develop transgenic resistant to Spodopotera litura and Heliothisis armigera |
| Rice Pusa basmati | *Cry1 Ab, Cry1 Ac* | For resistance against yellow stem borer |
| Mustard | *Ssu-maize Psy* and *Ssu-tpCrt1* gene | To generate plants, containing high levels of β-carotene |
| Rice | *GNA* gene | To generate plants, resistant to pests gall midge |
| Muskmelon | *Rabies glycoprotein* gene | To develop edible vaccines |

**Table 1.1** ***Contd.....***

## (b) Private Sector

| | | |
|---|---|---|
| Cotton * | *Cry1A(c)* | To generate plants, resistant to lepidopteran pests |
| Rice | *Cry1A(b), cry9C* and *bar* genes | To develop plants, resistant to lepidopteran pests and herbicide tolerance |
| Tomato | *Alfalfa glucanase* and *Tomato leaf curl virus* genes | To generate plants, resistant to viral and fungal attacks |
| Cotton* | *Cry1A (c)* | To generate plants, resistant to lepidopeteran pests |
| Cotton* | *CP4 EPSPS* | To generate plants, resistance to herbicide glyphosate |
| Cotton* | *CryX* gene | To generate plants, resistant to lepidopteran pests |
| Corn* | *Cry1A(b)* | To generate plants, resistant to lepidopteran pest |
| Pigeonpea | *GUS* | For transformation work |
| Rice | *Cry1A(c), Xa21 GNA* genes | To generate plants, resistant to lepidopteran pests, bacterial blight and sucking pests |
| Mustard | *CP4 EPSPS* | To generate plants, resistant to herbicide blight |
| Rice | *Bacterial blight resistance* conferring Xa-21 gene | To generate plants, resistant to bacterial blight |
| Brassica | *Bar, Barnase* | To develop superior hybrid cultivars |
| Mustard | *Barstar* | |
| Tomato* | *Cry1A(b)* | To generate plants, resistant to lepidopteran pests |
| Brinjal* | *Cry1A(b)* | To generate plants, resistant to lepidopteran pests |

**Table 1.1** ***Contd.....***

| | | |
|---|---|---|
| Cauliflower* | *Cry1H/cry9C* | To generate plant, resistant, to lepidopteran pests |
| Cauliflower* | *Bar, barnase, barstar* | To develop superior hybrid cultivars |
| Cabbage* | *Cry1H/cry9C* | To generate plants, resistant to lepidopteran pests |
| Cotton* | *Vip-3* gene | To generate plants, resistant to lepidopteran pests |
| Maize* | *Cry1A(b)* | To generate plants, resistant to lepidopteran pests |

*Transgenics developed elsewhere are being backcrossed to inbred lines to develop commercially viable hybrids that can be grown in different agro climatic regions of the country.

Source : Manju Sharma, et al, Agricultural Biotechnology in India, Status and Policy, Current Science, Vol. 84, Nos. 3.

## Agricultural Extension and Biotechnology

The transition from a resource based to a technology based system of agriculture, however places a greater responsibility on the agricultural extension sector since it is a vital conduit to transform and educate farmers on the new agricultural information and technologies as well as a conduit to research and policy makers on farmers problems, needs and concerns. It is now time that Agricultural extension should play a major role in public awareness and education of the farmers about myths, apprehensions, realities and approaches about biotechnology. Furthermore, it can be used as a vital policy instrument to promote participation among various stakeholders and to stimulate sustainable agricultural development.

Biotechnology extension, therefore, should be demand driven with focus on information embedded in inputs and markets: target groups as consumers and farmers with addressing concerns on profit making and protection of natural resources and environment. Both private and public extension services have equally important role in promoting biotechnology since there is need to communicate the information embodied in these technologies vis-a-vis substractablity and excludability.

## Challenges

The following challenges need attention of all concerned in order to accelerate the pace of development in the field of biotechnology.

- Do we have required database on cost-benefits of agricultural biotechnologies for undertaking public awareness campaigns ?
- Is it possible to take up collaborative research programmes involving private and public sectors? If so how?
- Do our scientists have enabling environment to cope up with the demands of recent WTO and GATT?
- Can the fruits of biotechnology be available equitably to farmers, local entrepreneurs and big companies?
- Can we afford to supply the improved products/technology as per the demand? If yes, what strategies we can adopt to promote them?
- How can we integrate Biotechnologies in the existing rural development programmes?

In order to translate the vision into action there is need for (a) accelerating the pace of research in biotechnology, and (b) transferring the technology to the ultimate users. This needs revitalization of research, extension and client system. It is widely recognized that growth and adoption of agricultural biotechnologies are hampered by inadequate public awareness about the strengths, weaknesses and potentialities of these technologies. A particular biotechnological innovation may be highly appropriate in one region but less suitable in another region. Similarly, this may benefit only a particular target group. This is because of the fact that some technologies are perceived costly and risky. Similarly, the rate of acceptance and adoption of even simple technologies are also not upto the mark, since there is lack of proper understanding of the scientific rationale and benefits. Besides, there are complex and expensive biotechnologies like GMOs which are subject of wide spread criticism as for as meeting the equity and biosafety concerns. The ongoing debate about transgenic crops is one such example in this direction. Thus there is varying degrees of perceptions, understanding, acceptance and adoption among various target groups. By and large, the degree of awareness of biotechnologies is very low.

## The project

Realizing the need for public awareness on and popularization of biotechnology for sustainable agricultural development, a social action research project has been undertaken by PRDIS with twofold objectives. This project is supported by BTU (APNLBP).

(a) To create awareness on the pros and cons of agricultural biotechnologies among all stakeholders at all levels in the project areas (and beyond) to enable them to take rational decisions.

(b) To develop and standardize the methodologies (methods, aids and strategies) for creating public awareness as well as to popularize biotechnologies through planned, systematic and sustainable communication campaigns.

During this process a series of workshops were organized for different target groups namely scientists, policy makers extension service providers and farmers in order to educate them on pros and cons of biotechnologies. These forums were used for developing and testing the A.V. aids, media and strategies for undertaking public awareness campaigns on sustainable basis.

Among the series of workshops organized, workshop-wise objectives and themes were as follows :

## 1. National Workshop Strategies for Public awareness on Biotechnology for Sustainable Agriculture

**Objectives**

- To elicit the views of various stakeholders on the application of biotechnology in agriculture.
- To discuss on the pros and cons of application of biotechnology in agriculture with focus on resource poor farmers.
- To identify the target groups, biotechnological interventions, methods (aids) and methodology to undertake sustainable public awareness campaigns in the rural areas.
- To evolve strategies and an action plan for undertaking public awareness campaigns.

**Themes**

- Social and economic issues (cost-benefits, equity and ethical aspects)
- Biosafety issues (environment, food, health and nutrition)
- Intellectual Property Rights (WTO, GATT, Patent right, Private and Public sector research enterprises)
- Communication strategies for sustainable public awareness (methods and media, target groups, technologies and sustainability)

**Target groups :** *Scientists, Extension Service Providers, Policymakers and Farmers*

## 2. Biotechnology for Sustainable Agriculture – Opportunities and Constraints

### Objectives

- To elicit the views of a multidisciplinary group scientists on opportunities and constraints in application of biotechnology for sustainable agriculture.
- To discuss pros and cons of application of different forms of biotechnology in agriculture with focus on higher forms of biotechnology.
- To identify strategies for: (a) public and private sector collaboration (b) strengthening research–extension-farmer–market-consumer linkages, and (c) pre and post release monitoring and management of genetically modified organisms (GMOs).
- To suggest communication strategies and messages for awareness in the scientific community.

### Themes

- Pros and cons of various forms of biotechnology interventions, particularly those relating to higher forms of biotechnology.
- Collaboration between scientists of public and private sector Intellectual Property Rights.
- Strengthening linkages among research, extension, farmers, markets and consumers in the field of biotechnology.
- Strategies for pre and post release monitoring and management of genetically modified organisms (GMOs)
- Communication strategies for improving awareness of biotechnology in the scientific community.

**Target Group:** *Scientists*

## 3. Biotechnology and Communication Strategies for Sustainable Agriculture

### Objectives

- To elicit the views of Extension Service Providers on the application of biotechnology in agriculture.
- To discuss pros and cons of application of different forms of biotechnology in agriculture with focus on resource poor farmers.
- To identify the biotechnological interventions, methods (aids) and methodology to undertake sustainable public awareness campaigns in the rural areas.

### Themes

- Lower forms of biotechnologies (vermicompost, biofertilizers and biopesticides)
- Intermediate forms of biotechnologies (animal vaccines and plant tissue culture)
- Higher forms of biotechnologies (genetically modified plants and animals)
- Communication strategies (methods and media)

**Target Groups :** *Extension Service Providers*

## 4. Biotechnology for Sustainable Agriculture – Challenges and Opportunities

### Objectives

- To discuss available biotechnologies
- To find out the advantages of agricultural biotechnologies and its problems in application
- To evolve communication strategies for creating public awareness

**Themes**

- Lower forms and Intermediary forms of biotech products
- Issues on genetically modified crops
- Communication strategies for sustainable public awareness

**Target Group :** *Farmers*

## 5. Biotechnology for Sustainable Agriculture - The way forward

**Objectives**

- To provide exposure on potential benefits and existing experiences of agricultural biotechnology applications.
- To discuss issues on biosafety and intellectual property rights and other concerns of people.
- To discuss the role of policy makers and administrators in promotion of agricultural biotechnology.
- To inform the policy makers about (a) recommendations made by previous public awareness workshop, and (b) suggested action plan for promoting biotechnology awareness in future.

**Themes**

- Importance of agricultural biotechnology for sustainable agriculture: experience of applying lower and intermediate forms of biotechnology.
- Modern Biotechnology in Agriculture.
- Recent trends in the development of agricultural biotechnologies
- Myths, apprehensions and realities in application of modern biotechnologies (e.g., Bt Cotton)
- Lessons from application of modern biotechnologies in different countries

- Policy issues in Agricultural Biotechnology
- Recommendations and Action Plan

**Target Groups:** *Policy Makers*

## Perspective Action Plan

Based on recommendations evolved during series of workshops, a perspective action plan was prepared involving key stakeholders for launching comprehensive public awareness campaigns in an effective manner and placed before the representatives of (experts) various institutions/organizations, for discussion and suggestions.

The issues raised through different themes during series of workshops contributed valuable articles. Of them some of the articles based on their importance, content and utility were selected and presented in four units : Importance of Biotechnology for Sustainable Agriculture, Different Forms of Biotechnology, Issues in Biotech Applications, Partnership and Communication Strategies.

## Bibliography

1. Annual Report, World Bank, 1999, Washington, DC, U. S.A.
2. Biotechnology Policy, 2001, Andhra Pradesh beyond tomorrow, Department of Industries and Commerce, Govt. of A.P.
3. Biotechnology News, January – March 2001, Biotechnology Unit, Institute of Public Enterprise, Hyderabad
4. Global Biotechnology Review, January 2, 2001.
5. Manju Sharma, Charak K.S. and Ramanaiah T.V. 2003, Agricultural Biotechnology Research in India - Status and Policy. Current Science, Volume 84, No 3. PP. 297-302
6. Norman E. Borlaug, 2000, Ending world hunger : The promise of biotechnology and the threat of anti-science zealotry, Plant Physiology, October 2000.
7. TERI, 2001, Global Status of genetically modified plants, **http://www.teriin.org/gmp/status.htm**

# 2

# Agricultural Biotechnology in the Present Indian Context

*M.V. Rao

## Trends in Indian Agriculture

- In the last 30 years, the net area sown under different crops in the country has remained more or less static at about 142 million ha. It would be difficult to bring more cultivable land, unless we reclaim wastelands, fallow lands, encroach on forestlands etc. As the population is rising, there is continuous decline in per capita arable land.
- There appears to be stagnation in the production and productivity of many of our crops. Productivity of many of crops is below world averages.

* Chairman, Andhra Pradesh Netherland Biotechnology Programme, Institute of Public Eneterprises, Hyderabad

Lead paper presented during the workshop for Policy Makers on Biotechnology for Sustainable Agriculture - The way forward.

- Production of coarse cereals, pulses, oilseeds and other crops grown under rainfed conditions is not moving up. In fact, there is drastic reduction in the area under coarse cereals like sorghum, pearl millet, and barley.
- Due to misuse or over use of water in some of our agricultural lands salinity problems are coming up which is reflecting in our sustainability of production system. Thus there is degradation in our soil-water systems.
- In spite of food grain production going up four times in the last five decades, still more than 260 million people in the country are below the poverty line. Many children and pregnant women suffer from malnutrition and vitamin deficiencies.
- Most of our crops have serious problems with several biotic stresses which involve pests, diseases, weeds as well as abiotic stresses like drought and salinity/alkalinity. In some cases, floods, cyclones, soil erosion and submergence destroy our crops. Some of these biotic and abiotic stresses are evading solution by the conventional crop improvement methods.
- The contribution of agriculture to the NGDP which was over 42 per cent in the fifties has steadily come down now to less than 25 per cent and so also the investment in agriculture.
- The biggest problem is the population, which has risen from 36.1 crores at the time of Independence, to more than 100 crores. All advances made in the food front are nullified by this population monster.

## Scope of Biotechnology

To meet the above challenges one has to "think" about the alternative routes besides exploiting the time tested conventional methods.

Biotechnology, based on the recent worldwide experience has already exhibited its efficacy and utility in the following areas:

- Herbicide resistance
- Pest and disease resistance
- Soil improvement
- Drought and salt resistance
- Nutritional improvement
- Yield improvement
- Shelf life improvement
- Environmental protection
- Bioconservation

## Global status of transgenic crops

Through genetic engineering and sophisticated rDNA technologies a number of transgenic crops are being produced in the world. Credit goes to the Peoples Republic of China, which for the first time commercialized transgenic crop varieties in the early 1990s. This was followed by USA in 1994 in clearing transgenic crops for food use.

During the eight-year period from 1996 to 2003, global area of transgenic crops increased from 1.7 million hectares in 1996 to 67.7 million hectares in 2003, with an increasing proportion grown by developed countries. Almost one-third (30%) of the global transgenic crop area of 67.7 million hectares in 2003, equivalent to over 20 million hectares, was grown in developing countries where growth continued to be strong. It is noteworthy that the absolute growth in GM crop area between 2002 and 2003 was almost the same in developing countries (4.4 million hectares) and industrial countries (4.6 million hectares), with the percentage growth more than twice as high (28%) in the developing countries of the South compared with the industrial countries of the North (11%).

For the first time India started growing GM cotton carrying Bt gene from the soil organism Bacillus thuringiensis. The most important crops in the exploitation of GM varieties are soybean, maize, cotton, canola followed by vegetables like cabbage, cauliflower, tomato, ornamental plants and fruits like straw berries, papaya etc. In 2003, the global market value of GM crops is estimated to be \$4.50 to \$4.75 billion, having increased from \$4.0 billion in 2002 when it represented 15 per cent of the \$31 billion global crop protection market and 13 per cent of the \$30 billion global commercial seed market. The market value of the global transgenic crop market is based on the sale price of transgenic seed plus any technology fees that apply. The global value of the GM crop market is projected at \$5 billion or more, for 2005. Besides reducing losses from specific pests/ diseases/viruses, the GM crops greatly helped in reducing pesticide application and in that way helped in reducing cost of cultivation to the farmer and also in reducing pollution of the atmosphere, soil and water bodies through the use of pesticides. Globally, it is felt that there is tremendous scope for biotechnology.

## The Indian Biotech Scenario

Government of India realizing the importance and scope of biotechnology, initiated a full fledged Department of Biotechnology (DBT) in 1986. The DBT encouraged, and is encouraging a number of research and developmental projects in several public and private institutions. The Indian Council of Agricultural Research (ICAR) established a National Research Centre for Biotechnology. Various National Institutions of ICAR have initiated biotechnology research to produce transgenic crops. A number of traditional universities have undertaken biotechnology and made significant strides in basic biotechnology, genomics, gene mapping etc. India is fast emerging as a technology driven country. The private sector, both national and transnational level realized the importance of biotechnology in the Indian agricultural scenario and are investing heavily in the sector. Field trials are being conducted with GM crop varieties of bell pepper, rapeseed, mustard, brinjal, cabbage, cauliflower, chillies, potato, tomato, tobacco and rice.

Several State Governments such as those in Andhra Pradesh, Karnataka, Tamilnadu, Maharashtra and Delhi are giving support to entrepreneurs to start biotech business in their States. They announced biotechnology policies and are in the process of setting up "Biotechnology Parks". The private sector is being encouraged to set up biotech industries in agriculture, pharmaceuticals, drug manufacture, vaccine producing etc. The Government of Andhra Pradesh set up a Biotechnology Park on 160 acres of land with Rs. 50.00 crores as initial capital and declared an area of about 600 sq.kms. as the 'Genome valley' to promote biotechnology. With this type of progressive steps by different State Governments and Central Government, it is expected that in the near future India may emerge as a biotechnology driven country.

The Department of Biotechnology, Government of India released a document "Biotechnology – A Vision" in 2001 outlining the goals for achieving in 10 years a sizeable and sustainable biotechnology industry. It also developed protocols and procedures for testing and commercial release of transgenic crop varieties and technologies.

## The Andhra Pradesh Netherland Biotechnology Programme (APNLBP)

The (APNLBP) programme started in 1996 with the assistance given by the Government of the Netherland to the tune of 44.40 crores. The Netherlands Government is supporting similar biotechnology programmes in Columbia (Latin America), Zimbabwe and Kenya (Africa), besides the Programme in Andhra Pradesh.

The main objectives of the programme are

- To promote application of biotechnologies relevant to small scale agricultural producers and processors in Andhra Pradesh.
- To develop appropriate biotechnologies through research activities on identified priority problems.
- To conduct supportive activities for development and adoption of biotechnologies including training, transfer of technologies and information dissemination.
- To strengthen the capacities of local organizations to develop and transfer biotechnologies and analysis of technology assessment.
- To promote the adoption of biosafety measures and discussion on IPRs.

The programme followed an 'interactive bottom up' (IBU) approach in prioritizing the felt needs of endusers, through interaction between farming community and scientists.

## Issues in Biotechnology

The following issues are realized to be critical for the success of biotechnology research:

- Increase investments in R & D in public sector institutions
- Prioritize research areas with focus on small farmers interest
- Promote public private partnership
- Create facilities for manpower development
- Ensure speedy clearances for the release of technologies, patents etc.
- Create facilities for risk assessment and safety assurance
- Systematic campaign for creating public awareness

When all these issues are addressed simultaneously and projects implemented on a "Mission-mode approach" it is expected that the goals would be realized in the specified time period.

While realizing the potentialities that are unleashed by biotechnology there is also a realization about the following risks and concerns:

- Creation of "Superweeds" and "Super pathogens and pests"
- Toxicity and allergenicity to humans and animals
- Erosion of biological diversity
- Socio-economic aspects
- Ethical aspects
- Trade related aspects
- Intellectual Property Rights (IPR)

The population pressure, food needs, sustainability of agriculture and ecosystem, pollution problems and emerging global scenario in the light of WTO, India has to play its own role in promoting biotechnology and be competitive in terms of economics, product development and quality in the international scenario. The country has also to take note of European and other countries with which it has export commitments, their objections and reservation about GM crops, foods and other biotechnologically development products. Biotechnology is not a magic wand to solve all our agricultural problems. This has to be kept in our mind that it is only one of the tools in the hands of plant improvement specialists, as opined by Ismail Seragaldin of the CGIAR (2001). "Agricultural biotechnology is not a magic wand that can replace poverty and hunger with a regime of plenty, but all available evidence suggest that it can be an effective additional weapon on the development front". Biotechnology results should be considered case by case and, they should be need based, before they are cleared. Environmental, social, political, ethical and economic issues should be carefully analyzed before we get biotechnologies released for commercial exploitation. Still there is considerable yield gap in most of our crops and animals, which can be bridged even by the conventional crop improvement method. Biotechnology could be essentially used to solve some problems, which are intractable and are not amendable for correction by conventional methods. This calls for considerable creation of awareness among the public and political leaders who are carried away by the biotechnology bandwagon projected by vested interests. In this context, the present workshop aims to bring such awareness among the organizers, hence need to be congratulated. I conclude by quoting Dr. Norman E. Borlaug, Nobel Laureate and Father of the Green Revolutions round the World:

**"Responsible biotechnology is not the enemy, starvation is. Without adequate food supplies at affordable price, we cannot expect world health or peace"**

# 3

# Biotechnology for Sustainable Agriculture Challenges and Oppurtunities

*William D. Dar

Biotechnology is becoming very popular, yet misunderstood and controversial. Biotechnology is becoming popular because of its tremendous potential in helping solve poverty, hunger, and environmental degradation in the developing world.

## The challenge

Notwithstanding its tremendous potential, the challenges faced by biotechnology are immense.

At present, more than 800 million people in the world are poor and malnourished. They live on less than a dollar a day. They are not sure that

* Director General, International Crops Research Institute for the Semi-arid Tropics (ICRISAT), Patancheru, Andhra Pradesh, India.

Inaugural address delivered during the workshop for scientists on Biotechnology for sustainable agriculture-opportunities and constraints.

their fields will yield enough food or that they will earn enough money to buy food. Forty years ago, there were about a billion hungry people. Two decades from now, there may still be 600 million poor people when world population will have ballooned to 8 billion.

Forty thousand people die each day of malnutrition, one-half of them children. Moreover, more than half a million children go blind each year from lack of vitamin A. Likewise, iron deficiencies are responsible for anemia among many millions of women and children, making them vulnerable to a host of diseases.

Sadly, even if food production was doubled by the Green Revolution, it did not solve the problems of malnutrition and hunger. Much more, it did not wipe out poverty. It did not create enough jobs to generate income that enables farmers to improve and sustain their livelihood.

In fact, poverty and hunger are the root causes of the global social conflict being experienced today. Esteemed Scientist, Prof. M.S. Swaminathan says, "where hunger rules, peace cannot prevail".

It is ironic that hunger persists while food prices are at an all time low. As the other half of the world lives in comfort and abundance, this situation is definitely not acceptable.

To face this challenge, the scientists should apply modern biotechnology especially those directed towards the poor. This should be combined with approaches to sustainable agriculture and greater participation of farmers in farm production, postproduction, and marketing.

## Revisiting biotechnology

A colleague at the International Institute of Tropical Agriculture best describes biotechnology as "a set of tools that allows plant breeders to introduce a greater array of novelty into their plant varieties, and select which works much faster than they could using conventional plant breeding techniques."

There is a wide range of areas where modern biotechnology can help improve the human condition :

1. ***Functional genomics :*** the molecular characterization of all the genes and gene products of a species.
2. ***Bioinformatics :*** the assembly of data from genomic analysis into accessible and usable forms.
3. ***Transformation :*** the introduction of single genes conferring potentially useful traits into plants, livestock, fish, and tree species.
4. ***Molecular breeding :*** the identification and evaluation of useful traits in breeding programs through marker assisted selection for plants, trees, animals, and fish.
5. ***Diagnostics :*** the more accurate and quicker identification of pathogens by using new diagnostics based on molecular characterization of pathogens.
6. ***Vaccine technology :*** the use of modern immunology to develop recombinant DNA vaccines for improved control against lethal diseases.

## Opportunities

Indeed, biotechnology offers enormous opportunities to practice sustainable agriculture, attain global food security, enhance human health, and improve the lives of poor people and farmers.

That is why this new millennium has been dubbed as a "Biotech Century".

Biotechnology can develop improved crop varieties that give higher yields with fewer inputs for our farmers. These crops can be grown in a wider range of environments, help conserve natural resources, yield more nutritious products with longer storage, and purchased at a lower cost by consumers.

For instance, these improved crop varieties contain insect resistance like that in cotton and corn, herbicide resistance in soybean, and delayed fruit ripening in tomato. Likewise, these improved crop varieties have better weed and insect control, higher productivity, and require more flexible crop management. This could prevent global crop production losses to as much as 35 percent, worth more than US $200 billion every year. These benefits accrue primarily to small farmers and agribusinesses. Economic benefits also accrue to consumers through lower prices.

At the same time, biotechnology can help farmers produce more nutritious crops. It can increase natural substances in plants like anti-cancer compounds soybean, vitamin A in rice, iron in cereals, or more non-saturated fatty acids in mustard oil.

Biotechnology also sustains the land's ability to support continued farming. This is done by developing crops that absorb nutrients more efficiently from the soil.

Plants can also be used to develop edible vaccines. Research is being done on fruits and vegetables that could one day deliver life-saving vaccines. For instance, banana is being developed that could soon provide the Hepatitis B vaccine. Likewise potato is being studied that could provide immunization against the Norwalk virus. This makes it possible to inoculate against deadly diseases with local and grown crops that are easy to handle, distribute and administer.

## The status of biotechnology

In 2001, the annual global review of commercialized transgenic crops reported the following :

1. The estimated global area of transgenic or GM crops was 52.6 million hectares, grown by 5.5 million farmers in 13 countries.
2. From 1996 to 2001, the global area of transgenic crops increased more than thirty-fold, from 1.7 million to 52.6 million hectares.

3. Four principal countries grew 99 percent of the global transgenic crop area. The USA grew 35.7 million hectares (68% of global total), followed by Argentina with 11.8 million hectares (22%), Canada 3.2 million hectares (6%) and China 1.5 million hectares (3%).
4. Globally, the principal GM crops were soybean planted to 33.3 million hectares, followed by corn at 9.8 million hectares, cotton at 6.8 hectares, and canola at 2.7 million hectares.
5. Herbicide tolerance has been consistently the dominant trait of GM crops followed by insect resistance.
6. The two dominant GM crops were herbicide tolerant soybean planted on 33.3 million hectares and Bt maize planted on 5.9 million hectares in six countries. The other GM crops occupied 5 percentage or less of global transgenic crop area.
7. On global scale, 46 percentage of the total soybean area were planted to GM crops; 20 percentage of the cotton area were GM; and 7 per cent the total area for corn. If the combined hectarage of these three principal GM crops is put together, this would constitute about one fourth of the total cropped area.
8. The number of farmers that benefited from GM crops increased to 5.5 million. More than three quarters of the farmers that benefited from GM crops were poor – planting Bt cotton, mainly in China and South Africa.

## Risks and constraints

Biotechnology offers not only the tremendous benefits and opportunities but also potential risks and constraints.

To make informed decisions on the appropriateness of biotechnology in relation to sustainable agriculture, food security, and human health, the following steps are needed :

(a) Carry out rigorous risk assessment and risk management within an effective national regulatory framework and

(b) Consider intellectual property rights to reward innovators and legally enable access to technologies developed by others.

Recently, intellectual property rights have become important in plant breeding, primarily through the greater use of utility patents. Such patents have stimulated greater research investment in industrialized countries. However, they also pose constraints to public sector breeding programs that produce seeds for poor farmers.

In assessing risks and benefits in the use of biotechnology, Gabrielle Persley suggest that distinguish between technology-inherent risks and technology-transcending risks.

Technology-inherent risks are those associated with food safety and the behavior of biotechnology based products in the environment. Technology-transcending risks emanate from the political and social context in using the technology and how this may benefit or harm user interests.

In spite of these potential risks and constraints, James & Krattiger in 1996 reported no risks associated with genetically modified crops in the last 14 years. This covered 30,000 field trials and hundred of millions of hectares of cultivated land.

Due to lack of public awareness about biotechnology, hostility to GM crops and foods is rapidly spreading from Europe to the developing world. Underlying this hostility is the concern on the lack of consumer benefits, business ethics, the environment, and potential impact on human health. These concerns must be genuinely addressed by national agricultural research systems and policy makers. Towards this end, field tests must be rigorously conducted in developing countries to assess the risks of using GM crops which may differ from those in industrialized countries.

## Biotechnology at ICRISAT

ICRISAT stands for the International Research Institute for the Semi-Arid Tropics, a non–profit, apolitical, international organization for science based agricultural development. Established in 1972, it is one of 16 centers of the Consultative Group for International Agricultural Research (CGIAR).

Harnessing Biotechnology for the poor is one of six global research themes, currently developing in line with new vision and strategy.

Under this theme, emphasis is on biotechnological approaches to crop improvement especially for traits most relevant to poor farmers and consumers in the dry tropics. Hence there are greater opportunities to utilize available genetic diversity. Main concerns are on abiotic and biotic stress traits including nutritional traits. Most activities will be carried out here in Asia. Scientists have long been involved in developing tools of molecular biology for the mandate crops of the institute, namely: chickpea, groundnut, pearl millet, pigeon pea and sorghum. Through these efforts, ICRISAT is in a position to apply biotechnologies in developing cultivars, which can tolerate drought, resist major diseases and insect pests. For example, usable molecular marker-based genetic linkage maps of chickpea, pearl millet and sorghum are available using these too many gene blocks that can reduce the vulnerability of these crops to biotic and abiotic stresses.

To attain global food security and fight hunger and poverty, agriculture this century must be made more productive at no expense to the environment guided by sustainable development framework. Therefore, assure that the impact of agricultural research reach the hundreds of millions of poor farmers who have not benefited from the "Green Revolution" that biotechnology can help meet this objective. National agricultural research systems (including states) must invest in biotechnology research to help poor farmers and consumers. Moreover, the public and private sectors must collaborate to promote the utilization of biotechnology.

Science or more specifically biotechnology, should be a means to serve the poor, not an end in itself. Even if excellent research is done but it has no impact in improving the lives of poor, it should be treated as failure. Only by using biotechnology to help reduce poverty, malnutrition and environmental degradation, a big difference in human development can be made. Hence, the research programs should not only generate knowledge about biotechnology, but generate relevant knowledge to increase the income of poor farmers and improve their quality of life in a sustainable manner. This is the human face of the science. This should serve as the overarching theme of efforts, the paramount motive of endeavours.

# 4

# Biotechnology for Crop Improvement - Facets, Prospects and Challenges

***Mangala Rai**

According to FAO estimates, global food grain production must increase from the present level of 2 billion tons to three billion tons by 2030. Compounding the present food situation is the realization that the additional food grains will have to be produced from less land, with less water, less labour and less chemicals, and most importantly, without degrading the fragile resource base. Thus, one of the major challenges facing the world in 21$^{st}$ century is to achieve food and nutritional security for the rapidly increasing population without degrading the resource base.

Plant improvement, whether as a result of natural selection or the efforts of plant breeders, has always relied upon creating, evaluating and selecting

* Director General, ICAR & Secretary, DARE, Govt. of India, Indian Council of Agricultural Research, New Delhi-110 001.

Lead paper presented during workshop for policy makers on Biotechnology for Sustainable Agriculture-The way forward.

the right combinations of genes. During the 20[th] century, traditional plant breeding has brought about enormous increases in crop productivity. However, plant improvement by hybridization is often slow, and is restricted to a very small gene pool owing to natural barriers to crossability. Beginning in the early 1980s, advances in plant cell culture and genetic transformation have overcome these barriers by making it possible to transfer defined genes into the desired genetic background. The gene pools of organisms – plant, animal, bacterial or viral – has now become one, reflecting our ancient concept of '*Vasudhaiva kutumbakam*'. The 1980s also marked the beginning of the era of powerful DNA-based markers, which can be effectively used for diverse purposes, including fingerprinting, analysis of molecular diversity, gene mapping and marker-assisted selection.

In conjunction with conventional technologies, modern biotechnology offers powerful means for increased and sustained productivity, improved product diversification and utilization, decreased reliance on agrochemicals and other external inputs, and better conservation of genetic resources.

## Diverse facets of biotechnology

It must be appreciated that there exists a gradient of biotechnologies varying in degree of sophistication, complexity, stage of development, and most importantly, application. At one end of the gradient, there are *in vitro* technologies, and on the other end, there are advanced technologies such as molecular breeding, genetic engineering, and genomics. The latter three have particularly caught the attention of the scientists and policy makers worldwide, mainly due to their wide-ranging applications and immense potential.

## Molecular Breeding

The discovery of Restriction Fragment Length Polymorphism (RFLP) in the 1980s as a DNA profiling technique and further elucidation of its potential use in identification of loci affecting important plant traits opened new horizons in genetic studies. Later, with the discovery of polymerase chain reaction (PCR) technique, there has been an exponential increase in the identification and utilization of marker systems suitable for genetic analyses in both plants and animals.

Molecular (DNA-based) markers are available in abundance in various crop plants. Innovative techniques making use of these markers have facilitated unambiguous identification of genotypes, besides analysis of genetic variation reliably and efficiently at the DNA level. Besides their significant contribution to the basic understanding of genomes, DNA-based markers have shown their value in mapping agronomically important traits, including simply inherited as well as complex quantitative characters, in various crop plants. Recent successes in gene tagging, QTL (quantitative trait loci) mapping, gene pyramiding, and marker-assisted selection, have formed the foundation for 'precision breeding'.

By combining the QTL approach with backcrossing, useful genes that control quantitative traits have been identified in the germplasm of plants not adapted to agriculture and have successfully been transferred to advanced breeding lines. "Unlocking the genetic potential from the wild" using this approach will be a major tool for future crop improvement. Molecular analysis has also revealed that the germplasm of wild relatives, either still available in the wild or conserved in gene banks, is vastly broader than the narrow gene pools of cultivated plant species.

Attempts are being made worldwide to efficiently integrate molecular marker technology into existing plant breeding programs to allow researchers to access, transfer and combine genes at a rate and with a precision not previously possible. As breeders strive to increase the number and complexity of traits that must be incorporated into new cultivars, they have to develop appropriate identification and screening procedures. The simultaneous or sequential screening for multiple biotic and/or abiotic stresses may become impractical or even impossible using conventional procedures. DNA marker technology offers significant promise in this context.

Attempts are being made worldwide, including India, to efficiently integrate molecular marker technologies into plant breeding programmes to allow researchers to access, transfer and combine genes at a rate and with a precision not previously possible. Molecular marker-assisted selection has already brought in a remarkable improvement in breeding for several traits, particularly disease resistance and improvement of quality traits in crop plants. Researchers are also now making steady progress in overcoming the hurdles associated with improvement of complex and quantitatively inherited traits, such as abiotic stress tolerance and yield, using molecular markers.

In the recent years, Indian researchers have demonstrated their expertise in development and application of molecular markers with respect to fingerprinting of genetic resources, analysis of genetic diversity in diverse crop plants, genome mapping and marker-assisted selection, particularly in crops such as rice, wheat, maize and brassica. Molecular breeding in crop plants needs to be further intensified for effectively addressing various priority areas.

## Transgenics

Basic studies in biochemistry and molecular biology worldwide have generated significant information about the genetic control and molecular mechanisms underlying several key biosynthetic pathways. It is now possible to identify, isolate, mobilize and express genes of interest within and across plant and animal kingdoms and develop 'designer crops' with novel traits.

Developments in gene cloning, transformation and regeneration of crop species led to an enhanced capability in developing transgenic lines in a variety of plant species. Institutions, particularly in the developed world, are successful in obtaining transgenics for diverse target traits, including insect pest resistance, viral disease resistance, herbicide resistance and quality traits. Worldwide, the beneficial impact of plant biotechnology has been so far felt almost exclusively on crops of high economic importance or those offering possibilities for 'value-capture' such as soybean, maize, sunflower, rice and potato. Other species as well as certain other traits, important to developing countries like India, have not attracted the interest of the private sector. An integrated and intensive national programme by ICAR would help in addressing this important lacuna.

Advances in gene discovery and genomics have led to the identification of several novel genes that provide excellent opportunities for effectively tackling biotic/abiotic stresses, for enhancement of crop productivity, and for improvement of nutritional quality. These scientific advances can accelerate the 'upstream' germplasm enhancement/pre-breeding for eventual crop improvement through effective transgenic approaches. Transgenic strategies can also facilitate development of enhanced germplasm populations precisely tailored to the needs of end users for target traits that are of particular relevance to India. A sustained national research capacity will be necessary for incorporating the target genes into locally adapted, tested, and effective varieties/genetic backgrounds.

## Plant Genomics

Any organism, including crop plants, can now be examined in terms of its whole hereditary organization through study of expression and interaction of genes – a field that is broadly referred to as 'genomics'. In the global arena, the genomics of a wild crucifer *Arabidopsis thaliana,* and model crop plant, rice, have already provided a wealth of information. India has contributed to this global endeavour as one of the partners in the International Rice Genome Sequencing Project (IRGSP).

Genome sequencing and gene annotation can describe all the genes in a plant, but they do not define their precise function. The focus of genetic research has now shifted from high-throughput sequencing to elucidation of gene function. Functional Genomics seeks to analyze the cellular and organismal functions of the genes and the manner in which their products interact to maintain those functions. The basic objective of plant genomics will be to use genomic information in combination with bioinformatics to enhance the productivity, quality and sustainability of our food production systems.

India should soon develop required capabilities in plant genomics as there are both opportunities and challenges to face in the years to come. The diversity of agricultural species in this country with their rich history of genetic improvement, their extensive germplasm collections both *in situ* and *ex situ*, and the intimate associations with human history and culture would certainly bring in exciting possibilities. Sophisticated and comprehensive computational tools will be needed to integrate information regarding genotypic performance, pedigree relationships, and germplasm diversity so that genomic data can be interpreted in ways that are useful to agricultural scientists. Diverse genetic resources, including mutants, germplasm, near-isogenic lines, mapping populations, landraces and elite breeding lines, are crucial for successful application of genomics for crop improvement.

## Salient challenges for application of biotechnologies in crop improvement

While India has made significant progress in the field of molecular biology and biotechnology, there are still some serious limiting factors to research activities and the use of biotechnology in India. These include inadequate infrastructure, lack of skilled human resources and availability of research equipment and facilities. These need to be addressed in a given time-frame.

Rapid advances in development and commercialization of transgenic crops in recent years have led to considerable apprehensions and concerns about the safety of the genetically modified plants for human and animal health and environment. In order to keep much needed pace with the rapid developments taking place worldwide, it is essential that precise mechanisms are evolved to assess the biosafety and performance of the new genetically modified (GM) products before recommending them for large scale cultivation. India, like several other countries, has developed procedures and guidelines for the evaluation and release of transgenic crops. The process of biosafety regulation needs to be both efficient and fast.

Science-based risk assessment of transgenics is the only way to effectively address public concerns, to dispel unwarranted myths and fears surrounding genetically modified foods, and instill confidence among the consumers about the biological safety of products derived from transgenics. Research experiments need to be initiated in India to effectively address some of the major concerns related to transgenic biosafety, thereby facilitating the risk assessment process. The country must also build scientific expertise, infrastructure and resources for biosafety evaluation and post-release management of transgenics.

## A strategic approach towards plant biotechnology

If knowledge and innovation are essential to meet productivity requirements, it is important that such needs are reflected in national objectives and priorities for food and nutritional security. To effectively reap the benefits of modern biotechnologies, India must develop a strategic approach. Some of the important components of such an approach are highlighted below:

## Enhancing research capacity in critical components

The nation must develop self-reliance in critical components of biotechnology, including isolation of genes for desired traits, isolation of promoters for organ-tissue-specific expression of the trait of interest, development of efficient transformation vectors, regeneration and transformation protocols for crop plants of present and potential importance to the nation. Increased investments and intensified efforts in a network mode are necessary to address the above mentioned areas. These areas also assume considerable significance from the viewpoint of Intellectual Property Rights (IPRs), which if not addressed effectively could potentially hamper development and commercialization of transgenics by the public sector institutions in India in the future.

## Strengthening supporting disciplines

Biotechnology can complement, but will not supersede, underlying scientific capacity in vital areas such as plant breeding and seed science and technology. Indeed in situations where plant breeding capacity is weak and where the seed sector is under-developed, there is a possible danger of frustration at not being able to see the fruits of research translated into biotechnology products. Therefore, adequate attention is being given to build national capacity in complementary underpinning technologies and development of infrastructure, which are necessary to ensure the transfer of technologies from the laboratory to the end users.

## Human resource development

Biotechnology in general, and functional genomics in particular, are relatively new areas in biological science. Many of the modern techniques for analysis of gene expression at RNA (gene chips/microarrays), protein (proteomics) and secondary metabolites (metabolomics) level are capital-intensive and therefore are being used by only a handful of institutions in the country. Besides, new innovations are continually being made in the areas of functional genomics and transgenics. These include powerful software for computational analysis of functional associations following high-throughput genotyping and phenotyping of the segregating lines and mutant populations. Although transgenic research is relatively more progressed in India as compared to the functional genomics, a distinct need exists for capacity building and human resource development in specific areas of transgenic research, particularly development of appropriate genotype-independent transformation and regeneration protocols, directed transfer of genes, and selectable marker-free transgenic development.

It is imperative that the scientists are trained in advanced research laboratories of the world to assimilate these novel techniques and appropriately apply the same to meet the national goals. A major initiative has also been taken to train young and talented scientists in modern biotechnologies, including transgenics and functional genomics, in reputed institutions worldwide.

## Partnerships and networks in biotechnology

Strengthening of partnerships and networks, both formal and informal, among those concerned with developing and distributing biotechnology products, as well as those interested in developing and disseminating information about biotechnology is of paramount importance. Such linkages should be facilitated and fostered not only within the national context, but also at bilateral, regional and international levels. Networks are especially important at this stage of development of agricultural biotechnology as many research teams and policy groups in various countries are on a 'learning curve' in dealing with the many different facets of biotechnology.

India has had a long and highly productive relationship with the international agricultural research centers, such as CIMMYT, IRRI, ICRISAT and ICARDA. These partnerships need to be further strengthened so as to ensure the development of biotechnology appropriate to the specific problems confronting agriculture in the developing world.

In India, ICAR has recently began a major initiative in developing network programmes within the country in important areas of plant biotechnology, including transgenics, functional genomics, molecular breeding, viral diagnostics, etc. This is expected to enhance the institutional capacity and research outputs through effective interactions, and sharing of germplasm, technologies and expertise, besides avoiding the duplication of efforts and wastage of resources.

## Public awareness about agricultural biotechnology

As countries like India are moving forward in terms of commercialization and human consumption of food products from genetically modified plants, there is a challenge of educating the general public, including farmers, consumers, scientists, extension personnel, policy makers etc. about the potential benefits and constraints of biotechnology in agriculture. Apprehensions triggered by misconception that all products of genetic modification have a genetic feature that makes them undesirable can only be laid to rest through effective communication. Effective communication and transparency are two key elements for successful commercialization of any new technology, transgenics being no exception.

A participatory approach is immensely required if the biotechnological products are to be accepted by the farmers and the general public. The mainstream population should be provided reliable and relevant information about biotechnology applications without hype or false hope. Besides measurable parameters such as yield, fitness, invasiveness, gene flow, toxicity or allergenicity, there are societal factors that need to be given due importance. These include consumer choice, long-term agricultural policy, ethics and societal responsibility for future generations. Scientists, from both public and private sectors, involved in biotechnology R & D have the primary responsibility in keeping the communication channels open and effective.

Recent developments in the areas of molecular biology, genetic engineering, *in vitro* methodologies and crop transformation have opened up exciting and uncommon opportunities for crop improvement. The whole spectrum of gene technology offers powerful means to successfully address a wide range of problems in plant biology, pathology, breeding and crop improvement in general.

The coming decades will bring increasingly intense research in this field, in institutions across developed and developing countries. India possesses the scientific talent, human resources and the capability to exploit biotechnology for the benefit of society in general and farming community in particular. The scientists must translate the potential to field reality in the near future. The major challenge before scientists in India is to develop and effectively implement integrated national networks on important areas in plant biotechnology, including transgenics, functional genomics and molecular breeding, for the nation to reap the benefits of biotechnological applications in agriculture.

# 5

# Strategies for Public Awareness on Biotechnology for Sustainable Agriculture

*V. L. Chopra

Mounting stocks of food grains in public storehouses notwithstanding, the need of increasing agricultural productivity as a function of area, time and inputs still persits. The increased productivity will have to be achieved through means and mechanisms different from those that have been employed in the past. The procedures of the future will require to be protective of natural resource base and more sensitive to the requirements of social justice and gender equity. In this context, new technologies assume relevance and biotechnology claims a place in national endeavor towards sustainable agriculture.

* President, National Academy of Agricultural Science, Indian Agriculture Research Institute, New Delhi

Keynote address delivered during National Workshop– Strategies for Public Awareness on Biotechnology for Sustainable Agriculture

In spite of many advantages that recommend biotechnology applications, the developing world has been apprehensive of its large-scale use. Its adverse impact on ecology and environment have been feared by some, while others have questioned its benignness to consumer's health. In this debate, emotions and ideologies seems to have taken precedence over scientific analysis, public awareness about facts, as they emerged from current knowledge, is necessary for making informed judgment both at the level of policy formulation and consumer choice.

What kind of strategies will aid formulating awareness efforts with requisite objectivity? The following suggestions may be helpful.

1. Accept that no technology can be guaranteed as completely risk-free to every one at all times. Technology adoption or rejection should be guided by cost to benefit and benefit to possible risk consideration.
2. Divide issues into workable categories, e.g., scientific, socio-economic, and related to health, environment, ecology and ethics etc. Deal with awareness strategies for each separately.
3. Avoid pre-judgement. Whenever testing is possible and practicable, do it with a sound design and make judgments only on the basis of critical analyses.
4. Accept time as a critically important factor. Decisions will have to be taken in a reasonable time frame. Delaying decisions only complicates matters and opportunities can be lost.
5. The process of testing and decision making should be totally transparent.
6. Testing and analytical processing should be done by independent and credible bodies
7. All shades of stakeholders should be fully associated at all stages. This process should be totally participatory.

In addition to these points, some more valuable comments were made during the course of presentation which are summarized below:

- Existing approach of Agricultural Development based on green revolution technologies (e.g., application of heavy doses of agro-chemicals) is not sustainable in the long run. In this regard an example was cited on yield response to nutrient application in 1975-80 and 1985-90 periods.

  During 1975-80 periods, each kg of nutrient application produced 15.3 kg of grain per ha. in contrast, the same amount of nutrient application produced only 8.1 kg per ha during 1985-90. This shows that for producing the same quantity of grain, the farmer has to apply double the quantity of fertilizers.

  Similarly, high chemical pesticides application for pest control in crop like cotton is becoming a serious problem year after year.

  This shows that there is need to look for new technologies.

- Biotechnology can provide a pathway for sustainable agriculture development. As an example, studies have shown that the insecticide application was reduced by 70-100 per cent in case of transgenic cotton as compared to the control. The yield advantage of transgenic cotton was 40 per cent over control.

- The main concerns and fears about biotechnology arise from the following:
  - Negative impact on environment
  - Health related concerns
  - Ideological and cultural concerns
  - IPR related concerns
  - Widening divide between developed and developing countries

- According to a report of US National Academy of sciences:

  "No strict distinction exists between the health and environmental risks posed by plants genetically engineered through modern molecular techniques and those modified by conventional breeding practices".

- The expectations from efficient agriculture in future for conditions prevailing in India are the following:
  - Job creating
  - Reduced production cost to benefit both
    - Producer
    - Consumer
- Increased production must result from
  - Increased productivity level
  - Production schemes responsive to ecological and environmental concerns
  - Processes sensitive to requirements of equity and balanced regional development
- Stepping up production is a major challenge for
  - Meeting increasing demands
  - Faster rate of growth
  - Facing developments arising from newer trade regimes e.g., WTO.

# UNIT - II

## Different Forms of Biotechnology

# 6

# Biofertilizers

* P. Ranga Reddy

Biofertilizers are microbial preparations consisting of living cells of bacteria, fungi and algae alone or in combination which may help in increasing the crop productivity by way of helping in biological nitrogen fixation, solubilization of insoluble fertilizers compounds, stimulating plant growth or in decomposition of plant residues. In a larger sense, it may use to include all organic resources (manures) for plant growth, which are rendered in an available form for plant absorption through microorganisms or microorganisms plant association or interactions.

* Scientist, Sri Aurobindo Institute of Rural Development, Gaddipalli, Nalagonda District, A.P.

## Biofertilizers may broadly classified into three groups

(a) Nitrogen biofertilizers-*Rhizobium, Azospirillum, Azatobacter, Blue Green Algae, Azolla.*

(b) Phosphorous Biofertilizers-Phosphobacterial organisms like *Pseudomonas* and *Bacillus.*

(c) Organic Matter decomposers-mediated biofertilizers like FYM, Green manure, Vermicompost.

## Nitrogen Biofertilizers

Fixation of atmospheric nitrogen for the crop growth is being carried out in two ways by different bacterial organisms. Species belonging to Rhizobium carry out the process through the formation of root nodules and it is called symbiotic nitrogen fixation. On the other hand, free living microorganisms like *Azotobacter* and *Azospirillum* fix nitrogen without any association with host plant system and is known as non-symbiotic nitrogen fixation.

***Rhizobium :*** *Rhizobium* is the best biofertilizer meant for legume crops including pulse legumes like green gram, red gram, pea, lentil, black gram and cowpea etc., oil seed legumes like groundnut and soybean, forage legumes like barseem and forest legumes like subabul, babul, sesham etc., it has the host specificity. It cannot form nodules or fix nitrogen across leguminous group of crops. As such, farmers have been advised to obtain crop-wise cultures of Rhizobium.

***Azotobacter and Azospirillum :*** Nitrogen fixation by these organisms has been reported to be around 20-40 Kg N/ha. when applied to a range of different crops like jowar, bajra, wheat, rice, maize, barley, oil seed crops, sugarcane, cotton and different horticultural and ornamental plants. They secrete and supply hormones and vitamins required for crop growth.

## Blue green algae

Rice cultivation with standing water encourages the growth of *blue green algae*. The *blue green algae* possesses twin abilities of photosynthesis as well as biological nitrogen fixation. Besides fixing nitrogen, these algae excrete Vitamin B12, auxins and ascorbic acid which may also contribute to the growth of the rice plants.

Farmers can prepare their own algal biofertilizers. It has been recommended that trays (6'X3'X9') of galvanized iron sheet or brick and mortar or pits lined with polythene sheets filled for 2"-6". When the soil settles, soil based starter culture is sprinkled over the water. Within a week, thick algal scum is formed on the surface. It is also recommended that such algal preparation is applied at the rate of 10 kg/ha in the field, one week after transplantation.

## Azolla

*Azolla* is a floating fresh water fern and is ubiquitous in distribution. It has an algal symbiotic (*Anabaena azolla*) within the central cavity. The algal fixes atmospheric nitrogen and is present at all stages of growth.

***Application :*** Two methods of *Azolla* application have been recommended in India;

(a) As a green manure by incorporating in the field prior to rice planting

(b) By growing *Azolla* along with rice.

Results indicated that incorporation of 10 tons of fresh *Azolla*/ha is as efficient as basal application of 25-30 kg. N/ha. The practice of incorporation of *Azolla* in soil appears to be more beneficial than dual culture. It is also feasible to use *Azolla* along with the application of nitrogen fertilizer to step up rice yields.

## Pseudomonas solubilising microorganisms : Pseudomonas and Bacillus

Next only to nitrogen, phosphorous is a vital element for plants and microorganisms. The inorganic form of the element in the soil are compounds of calcium, iron, aluminium and fluorine. The organic forms are compounds of phytin, phospholipids and nucleic acids which come mainly by way of decaying vegetation. Therefore, soils containing high organic matter are also rich in organic forms of phosphrous. Super phosphate is one of common forms of phosphatic fertilizers. Rock phosphate is one of the basic raw materials for phosphatic fertilizer production.

Several soil bacteria particularly those belonging to the genera of *Pseudomonas* and *Bacillus* possess the ability to bring insoluble phosphates in soil into soluble forms by secreting organic acids such as formic, acetic, propionic, lactic, glycolic, humic and succinic acids. These acids lower the $_{p}H$ and bring about the dissolution of bound forms of phosphates.

## Organic matter decomposers

Organic matter represents the remains of all kinds of plants, animals and microorganisms in various stages of decomposition. Soil microorganisms of different kinds do the actual work of breaking down the organic substances. Each kind of organisms plays some significant role in the decomposition of plant and animal residues, liberation of plant nutrients and in the development of soil structure. Fungi form an essential part of soil microflora. These organisms form tiny threads called mycelia, which enmesh soil particles into granules while the bacterial organisms perform most important function of breaking down organic matter. They are mainly nitrate working, cellulose destroying, sugar and starch working bacterial organisms.

## FYM (organic manures)

Organic materials are valuable by-products of farming and allied industries derived from plant and animal sources. They are bulky in nature but supply small quantities of plant nutrients, e.g., Farm Yard Manure (FYM), rural and town compost, green manures and night soil etc. However, there are others like oil cakes, fish meal and poultry manures etc., containing higher percentages of major nutrients like nitrogen, phosphours and potassium (NPK).

Compost materials are the decayed refuse like leaves, twigs, roots, stubbles, crop residue, water hyacinth, sawdust and street refuse. The process of decomposition is hastened by adding nitrogenous materials like cow dung, urine or fertilizer. Large number of microorganisms feed on these wastes and convert into well-decomposed manure. The final product is known as compost.

Farmyard manure improves soil properties including physical, chemical and biological characteristics. Application of organic manures would change the rhizosphere environment by affecting porosity, aeration, temperature, and water holding capacity and soil micro flora. They contain necessary ingredients like nitrogen, phosphorus, potassium, sulphur, calcium and magnesium for crop growth. Oil cakes are rich in nitrogen, phosphorus and potash. Oil cakes need to be well powdered before application to the soil for easy decomposition by microorganisms. In general, application of FYM or compost is the best source for maintenance of soil organic matter in Indian soils.

## Green manure

Green manure supplies organic matter to the soil. There are different kinds of green manure crops like glyricedia, sesbania, green gram, bean, cowpea and horse gram. They can be grown in the fallow soils and ploughed down into the soil just before flowering. Through this process, lot of organic matter in the form of leaves, tender twigs and roots are added to the soil. They get decomposed in the soil through the action of microorganisms converting into manure. It supplies nitrogen to the extent of 25-30 per cent per hectare. It also helps in improving soil structure and in converting the saline/alkaline soils into aerable soils good for crop growth.

## Vermicompost

Vermicompost is regarded as a valuable biofertilizer and an important component of farming package. It is also rich in plant nutrients, beneficial soil microbes, vitamins, antibiotics and plant hormones.

Vermiculture and vermicomposting are two interlinked and interdependent processes. Vermiculture can only be done on compostable or decomposable organic matter. Composting is the outcome of earthworm activities. So, both the processes can be brought about simultaneously.

## Mass production of bacterial inoculants

Biofertilizer production is based on batch culture technology in which the desirable bacteria is multiplied in the specified media in vessels varying from conical flasks to large vessels of about 1000 ML capacity. The following steps are required for production.

(a) Isolation and strain selection

(b) Preparation of mother culture

(c) Fermentation

(d) Preparation of carrier based inoculants

(a) **Isolation and strain selection**

An efficient strain of the microorganism is as important in the biofertilizer production as seed in crop cultivation. Choice of these strains is very critical task as the success of biofertilizer depends on its effectiveness in terms of N-fixation of solubilisation as the case may be.

(b) **Mother culture**

Cultures to be used for mother culture preparation are best maintained in freeze dried form in glass ampules. From the fresh culture, starter inoculums is prepared in 1000 ML conical flasks containing 250 ML medium. The flasks are shaken in rotary shaker for 3-7 days depending on the growth. This culture is then transferred to mother culture medium either in conical flask of 1000 ML capacity or fomenters depending on the requirement.

(c) **Fermentation**

Large quantity of liquid medium (broth is formulated in production fermentor and sterilized). The mother culture is transferred aseptically to the production fomenters. The factors influencing the output of bacterial cells are aeration, volume, initial inoculums level, bacterial strain, and temperature and incubation time. The prime objective is to attain high population in minimum time.

(d) **Preparation of carrier based inoculants**

Fermented broth having bacterial population of about 10 cells/ml is blended with carrier (pest or lignite). The mixture is left for curing for 2 days and then the product is packed in polythene bags.

## Application of technology for bacterial inoculants

### (a) Seed treatment

(i) Dissolve 50 gm of country sugar in half liter of water, heat for 15 minutes and cool it to room temperature.

(ii) Mix biofertilizer in it to form a slurry.

(iii) Add slurry to the seeds required to plant.

(iv) Mix the seed with the slurry in such a way that all seeds should get uniform coating of the inoculums.

(v) The treated seed is dried under shade for about 30 minutes, before sowing.

The mixture of *Rhizobium* and *Phosphobacterium* for legume crops and the mixture of *Azospirillum/Azatobacter* and *Phosphobacterium* can be used for all the other non-leguminous crops.

### (b) Seedling treatment

For transplanted crops like rice and some vegetables, the mixture of *Azospirillum* and *Phosphobacterium* can be used.

Mix 1 kg. of biofertilizer mixture inoculums in 10 lit. of water.

The roots of seedlings can be dipped in the suspension for 20-30 minutes before transplanting.

### (c) Soil application

The mixing of *Azospirillum* and *Phosphobacterium* can be mixed in 10-20 kg. of sand or fine compost.

This mixture can be directly applied to the soil.

This method is recommended for crops like sugarcane, vegetables and horticultural crops. Usually 1-2 kg of biofetilizers can be applied to one acre of field crop.

## Advantages and limitations of biofertilizers

### Advantages

(a) Biofertilizers can complement the use of chemical fertilizers. It can help deficiency in fertilizer consumption.

(b) They are capable of converting elemental form of nitrogen or the free atmospheric nitrogen into the form, which becomes readily available to the plant.

(c) Some biofertilizers solubilize and mineralize the phosphorus and increase uptake of phosphorus by the plants.

(d) Bio-inoculants also enhance plant growth by the release of hormones and auxins etc.,

(e) Their use increase crop yield by 10-20% on an average.

(f) Some biofertilizers protect the crop from various soil borne diseases by suppressing their effects with their antagonistic behavior against pathogens.

(g) They help to proliferate and survival of other beneficial microorganisms in soil.

(h) Improves soil properties and sustain soil fertility.

(i) Provide Plant nutrition at low cost and benefits the consecutive crops.

### Limitations

(a) Results are not as dramatic as with chemical fertilizers

(b) Depends upon the soil type and condition and moisture.

(c) All crops are not equally responsive to the application of biofetilizers.

(d) It is slow process-take time to boost the population of soil micro flora.

## Vermicomposting materials

(a) Dung obtained from different animals and poultry droppings.

(b) *Agriculture waste :* Plant parts like stem, leaves, husk, peels, vegetable waste, orchard leaflitter, processed food waste, sugarcane trash etc.

(c) *Forestry waste :* Various types of plant parts like wood shavings, peels, saw dust plup and forest leaflitter.

(d) City leaf litter

(e) Biogas slurry

(f) Waste paper and cotton cloth

## Steps in vermicompost preparation

(a) Vermi composting area should be free from stones, plastics and glass etc., and sufficient water sprinkled over the place.

(b) A 4-inch thick layer of agro waste like sugarcane trash, paddy straw, banana leaves and stalks of pigeon pea and cotton etc., is laid.

(c) After a layer of agro-waste a layer of half decomposed farmyard manure or biogas slurry should be placed to a height of 3". This layer acts as feed material for earthworms and maintains temperature by conserving moisture.

(d) A layer of earthworms should be equally distributed all over the bed at 2000 earth worms/meter.

(e) A layer of dung waste mixture (dung mixed with finely chopped agro waste) into a thickness of 12" forms main raw material for bioconversion.

(f) The next layer is fresh dung mixed with green matter (leaves, weeds) up to a height of 5-6 inches.

(g) The upper surface of the bed should be completely covered with gunny bags.

(h) The sides of bed should be lined with agro-waste to prevent outward movement of temperature.

(i) Water should be sprinkled on gunny bags according to moisture in the bed and atmospheric temperature.

(j) The bed should be examined every week to monitor the temperature pH and colour of vermicompost (ultimately black in colour)

(k) Vermicompost is ready within 4-5 weeks period.

(l) When vermicompost is ready, do not sprinkle water on the bed for 2 days to keep the upper layer dry. Earthworms will move down to the lower layer.

(m) Gather vermicompost into a heap of conical form so that the earthworms move to the lower layer of the heap.

## Advantage of vermicompost

1. Huge quantities of domestic, agricultural, rural organic wastes can be recycled for vermicompost preparation.
2. Vermicompost substitution with fertilizer will reduce economic input
3. It can be produced near the site of use.
4. Extra production with or without worms can be marketed for extra income.
5. Vermicompost is rich in several beneficial micro flora and also in certain enzymes and growth regulators.

6. During the process of earthworm's multiplication, they churn and tunnel into the soil to make it porous, improve water infiltration, moisture retention essential for rain fed areas.
7. Prevents soil degradation and enhance fertility status.
8. Reduce soil toxicity by buffering action.

## Limitations of vermicompost

1. Poultry birds and rats eat away earthworms.
2. Cattle population getting reduced with farmers and so availability of adequate dung may become constraint.
3. As a household activity it has to be sustained despite other works in the house.
4. Large quantity is required for application – minimum of one ton/ acre.

## Demonstrations in the farmer fields

Demonstrations were organized in 14 fields from neighboring villages on different crops to bring awareness to the farmers on the impact of biofertilizer application on the crop growth and pod yield. The crops are Cotton, Castor, Groundnut and Red gram from different villages. The unit supplied the biofertilizers and every farmer was advised to keep half of their field as untreated checks to know for themselves the comparative performance. The samples were collected at the time of harvesting from treated and untreated plots and calculated as to the yield per acre. The results clearly indicate that the increase in the yields varied from 11.7 to 24.6 percent for Rice and to an extent of 18.1 to 35 percent for Groundnut. It also shows that the combination of nitrogen fixers and Phosphate Solubilisers contribute greatly for overall increase in yields of different crops.

# 7

# Botanical Pesticides

* G. Srimannarayana

All over the world, one-third of the agricultural produce is being damaged by about 20,000 types of pests, revealed by the recent survey. Though the application of chemical pesticides played a major role in eradicating the crop pests, yet they could not give the optimum results. Indiscriminate use of chemical pesticides not only increased the resistance power of insect pests but also caused the atmospheric pollution. Further more these chemicals are also damaging the predator insects. The human health is also being affected due to the residues of the chemical pesticides.

In these days Integrated Pest Management (IPM) is gaining much importance. The main aim of this is to restrict the use of toxic chemical pesticides and to use the available botanical pesticides to control the crop

* Former Professor, Osmania University, Hyderabad

pests. The Integrated Pest Management occupies a vital role in the control of pests and diseases. It is a broad ecological approach which aims in keeping pest population below the economic threshold level by blending more than one method of pest control such as cultural, mechanical and biological methods in a compatible and environmentally sound manner.

Gradually, the farmers are getting acquainted with the usage of plant origin pesticides which are more eco-friendly in place of chemical pesticides which are not only polluting the atmosphere but also harmful to the human and animal health. In India, there are about 1744 plant species having the characteristics of killing the crop pests. In these, 1005 varieties are pest killers, 384 pest inhibitors, 238 repellents and 27 are attractants. Among these plant species, Neem, Seethaphal (*Anona Squamosa*), Pongamia are very important.

## Neem (*Azadirachta Indica*)

In pest management, the neem plant has become more important. In India, from the times immemorial, the neem plant is identified as having insecticidal and medicinal properties. The neem seed contains *Azadirachtin*, a chemical which is an insect inhibitor and stops the pest population growth. It also contains about 30 types of other limonoid chemicals. The Neem Seed Kernel Extraction (NSKE) and neem oil are having significant importance due to these chemicals. The NSKE and neem oil in combination with other chemicals are also available in the market. The farmers can themselves prepare the NSKE in which apart from *Azadirachtin*, there are other important chemicals like Selanin, Nimbin, etc., The neem seed can be powdered in a grinder. 50g of grounded powder can be mixed with one liter of water, packed in a muslin cloth and soaked for about five hours in water. Later the powder should be squeezed in the water completely. To this 5 per cent NSKE, some soap powder should be added and mixed and filtered it well. Thus the NSKE is ready for spraying. Commercially the collection of neem seed and manufacturing of the neem by-products has increased considerably. The neem products are abundantly available in the market in comparison to other botanical pesticides. The neem products are not only used as pesticides but also used in other products like fertilizers, toothpastes, soaps, beauty aids and mosquito repellents and in ayurvedic medicines.

## Sithaphal (*Anona squamosa*)

Sithaphal plant is most famous for its sweet and flavoured fruits but it is wonder to note that its seeds contain acetogeneces, the chemical, which kills the pests and also used as inhibitor. Similarly, the alcoloids contained in the leaves of this plant are also useful as the repellents of pests. The farmers can prepare the pesticides with the leaves as well as from the seeds. 50g of leaves can be soaked in one liter of water and the extract can be prepared. This can be prepared by two methods. One is to mix the leaves with water and groundwell and filtered. Then the same is filled in the sprayer and mixed with small quantity of soap powder and sprayed. In second method, 50g of leaves can be boiled in one litre of water and when the water becomes dark, coloured solution thus obtained can be cooled, and filtered. Now mixed with some soap powder and the extract is ready for spraying. The oil extracted from the seeds contain acetogenins which are more powerful. At farmers level, the seeds are powdered and socked for 12-24 hrs (50gm of powder in 1 lit of water) then squeezed to get extract and filter to spray the solution.

## Pongamia (*Pongamia glaba*)

Everybody knows that the leaves, branches of Pongamia can be used as green manure. Similarly, the pongamia seeds are useful for the pest management. In these seeds the available chemical is “Karengina”. This chemical is useful to kill the pests and as pest inhibitor. The powder of pogamia seed and pongamia seed oil can be used for pest management. 50g of seed powder is to be soaked in one litre of water for 12 hours and squeezed fully to get the extract and then a small quantity of soap powder is added. This is now ready for spraying. If the pogamia oil is available in the market, 5 ml of oil may be mixed with one litre of water and soap powder as required. This is ready for spraying. The ready to spray pogamia oil EC is also available in the market, which can be directly mixed in about 5 ml to one liter of water and sprayed.

## Other kinds of plant origin pesticides

Garlic, onion, chillies extract : 100g of each of these items can be boiled in water for about 1-2 minutes and finally grounded and then added to the water and sprayed on the crops.

## Tobboco Extraction

Tobacco leaves, stems or powder can be boiled in water for about 3-4 hours and the extraction is ready for spraying.

## When to spray the plant origin pesticides

The effect of these pesticides is lost due to sunlight. Therefore, these pesticides should be sprayed preferably in the evenings, so that the pest menace could be controlled effectively.

Botanical pesticides can be used for controlling tender skinned sucking pests and leaf eating caterpillars. As a preventive measure this can be sprayed on the crops, as a result butterflies and moths do not lay eggs and eggs do not hatch. First instars larvae do not feed on those leaves. It is difficult to control grown up caterpillars with these pesticides. Oily preparations lasts long but leaf extracts have short life. Hence leaf extracts should be sprayed more number of times and oily preparation should not be sprayed at high concentration or manytimes because it gives scorching effect and crops dry up. Leaf decoctions can be sprayed at 7-10 days intervals, whereas oily preparations can be used at 10-15 days intervals.

# 8

# Biopesticides

* P. S. Vimala Devi

The increasing pressure to produce more and more crops has resulted in wanton use of chemical pesticides resulting in criminal pollution of our ecosystem. This is amply reflected in the deteriorating health of people. As per the WHO estimates, 1 million cases of pesticide poisoning occur every year and 20,000 deaths globally. This damaging ecological disturbance is the outcome of injudicious use of pesticides. The chemical pesticides have come under attack globally thus casting on us a responsibility to adopt eco-friendly pest control measures. Biopesticides are the ultimate refuge in this scenario. The term biopesticide includes any biological material and organisms, which can be formulated as pesticides for the management of pests. Among biopesticides are botanicals such as extracts and natural actives - compounds of neem, pongamia etc., and microbial pesticides,

* Sr. Scientist, Directorate of Oilseeds Research, Rajendranagar, Hyderabad - 500 030.

which are whole organisms or their secretary/excretory products. The conservation and augmentation of these biopesticides are the natural control options to manage insect pests and plant diseases. Control through biopesticides is consistent with the goal of sustainable agriculture and IPM to minimize the use of chemical pesticides around 60 per cent of natural control of insects in any crop is through biocontrol agents, which must be protected and preserved by precluding needless application of chemical pesticides.

## Biopesticides

The last two decades have witnessed a radical change in pest management strategies with greater attention being paid to biological control (as a part of Integrated Pest Management). In India, the most commonly used microbial insecticides *Bacillus thuringiensis (Bt).* and *Bacillus subtilis (Bs)* were approved in 1993 for large scale field use in the country and seven agrochemical companies started marketing of the imported products. Neem usage in India for pest control is an age old practice. However, commercial production by the organized sector companies started in 1992 with the launch of Nimbecidine by SPIC. The market started growing after 1994-95 with the entry of more companies. Increasing demand for eco-friendly products as well as pressure from Government and environment bodies as well as possibility of good profit has rendered its inclusion in product portfolios. The effect of neem based pesticides is not as visible as with chemicals. However, its ability to manage resistance positions it is in supplementary role to chemical pesticides. Neem can become a basis for chemicals free, local resource based product if efforts are directed at educating the farmers, by providing training through Government and NGOs in preparation of various neem products.

The registration committee constituted under the Insecticides Act, 1968 has approved the registration of *Bt, Bs,* and allowed their commercialization during the period of their provisional registration. The other biopesticides approved by the Central Insecticides Board include NPV, GV, entomogenous fungi *(Beauveria bassiana, Metarhizium anisopliae, Verticillium lecanii and Nomuraea rileyi), antagonistic fungi and bacteria viz., Trichoderma spp. Gliocladium spp, Pseudomonas fluorescens and Bacillus subtilis, mychoherbicides,* insect repellants and pheromones to regulate their manufacture and use in the country. Globally, the commercial application of microbial pesticides is just one per cent of the total pesticides. Of these, *Bt* alone comprises 90 per cent of the total use of microbial pesticides.

## Advantages of Biopesticides

- Ecosafety, biodegradability, hence minimize environmental hazards
- Compatibility with beneficial biocontrol agents, self-perpetuating
- Host specific
- No development of resistance
- Cost-effective
- No pest resurgence
- No waiting time for harvest
- Enhance acceptability and value of produce for exports
- Well suited for rural areas

## Limitations of Biopesticides

- Slow acting
- Brief shelf life
- Time consuming preparation which is highly technical requiring more effort to make high production targets feasible
- Short span of actual demands
- Demand unstability such as cancellation of orders

**Table 8.1 Commercial products of *Trichoderma* available in India**

| Product/antagonist | Target pathogen/host | Source/reference |
|---|---|---|
| *Sun-derma (T.Viride)* | *Rhizoctonia solani Macrophomina* | Sun Agro Chemicals, Chennai, India |
| *Ecofit (T. Viride)* | *R.Solani, M. Phaseolina* | Hoechst Schering Agro Evo Ltd. Mumbai, India |
| *Trieco (T. viride)* | *Root rot and wilts* | Eco Sense Laboratories (I) P. Ltd, Mumbai |
| *Antagon – TV (T. Viride)* | *R. Solani, M. Phaseolina* | Green Tech Agroproducts (P) Ltd., Coimbatore, India |
| *Trico (T. Viride)* | *Several* | Pralshar Biotech (P) Ltd., Goa |
| *Biohit (Trichoderma Spp)* | *Several* | Indore Biotech Inputs & Research (P) Ltd., |
| *Biofit (T.Viride)* | *Several* | Ajay Biotech (I) Ltd., Pune |

Table 8.2 ***Bacillus Thuringiensis*** **formulation available in India**

| Product | Insecticide potency | Manufacturer/supplier | Variety of BT |
|---|---|---|---|
| Agree | NA | Hindustan Ciba Geigy Ltd | *Kurstaki* |
| *Bioasp* | *3200 Iu/mg* | Bio Tech International | *Kurstaki* |
| *Biolep, BTK-I* | *3000 Iu/mg* | | *Kurstaki* |
| *BTK-II, Lepidocide* | *3000 Iu/mg* | | |
| *BTT* | *1500 Ii/mg* | " | *Thuringiensis* |
| *Biotoxibacillin* | *3000 Iu/mg* | " | *Thuringiensis* |
| *Dendrobacillin* | *3000 Iu/mg* | " | *Dendrolimus* |
| *Biobit* | 32000 Iu/mg | Rallis India | *Kurstaki* |
| *Dipel 8L* | 17000/ml | Lupin India/Cheminova India Ltd. | |
| *Delfin* | 53,000 Su/mg | Syngenta India/Margo India | *Kurstaki* |
| *Halt* | 55,000 Iu/mg | Wockhardt Life Science | *Serovar Kurstaki (Indigenous)* |
| *Spicturin* | I × $10^6$/ml (spore count) | Tuticorin Alkali Chemicals | *Galleriae* |

# 9

# Plant Tissue Culture Technology

* B. Venkateswarlu

Tissue culture is a method of regenerating a whole plant from cells, tissues or organs under defined environment. This technology is based on 'totipotency' of plant cells i.e., an inherent ability of a cell to regenerate an entire plant if suitable media and growing conditions are provided. Started nearly a century ago, plant tissue culture technology made rapid advances during the last three decades. Today, it is extensively used in agriculture, horticulture, forestry and medicinal plantations. For several years, tissue culture technology remained confined to the laboratories. Of late, it is moving to the farmers' fields with large number of horticultural and forestry plants being multiplied commercially both in public and private sector and

* Principal Scientist, Central Research Institute for Dryland Agriculture, Santhoshnagar, Hyderabad - 500 059.

made available to the farmers. In addition to the propagation of true to type plants, tissue culture technique is also used in a number of basic and applied research fields.

## Types of Tissue Culture

Literally, tissue culture means regenerating a plant from a small piece of plant tissue. However in practice, this term has wider meaning. It includes use of single cells, protoplasts (cells without cell walls), tissues, meristems and whole organs to produce a complete plant. Following are the main types:

(a) **Cell and protoplast culture :** This refers to growing single cells or protoplasts in defined media and regenerating the whole plants. This method is useful in production of cytoplasmic hybrids and transgenic plants.

(b) **Anther culture :** This refers to the culture of whole anther or micropores to produce haploids. Haploids are useful in hybridization experiments. They are also used to produce homozygous diploids. Anther culture is extensively used in China in hybrid rice production.

(c) **Ovary culture:** Ovaries from pollinated flowers can be cultured in the medium to produce healthy fruits. However, it is used mostly for experimental purpose.

(d) **Embryo culture:** Embryos with wide cross hybrids can be rescued (saved) by culturing them in the laboratory. This is extensively used in grass/cereal family.

(e) **Meristem culture:** Actively growing shoot epicess are cultured in the laboratory to produce plants free from viruses. This is widely used in horticulture.

(f) **Tissue and organ culture :** When complete organs like seeds, leaves or nodal segments are used to regenerate a plant, it is called organ culture. Sometimes, parts of the organs like leaf discs are also used to regenerate whole plants. It is most commonly employed in genetic transformation or production of transgenic plants.

(g) **Micropropagation :** When a portion of the plant like leaf, somatic embryos or nodes are used to produce whole plants *in vitro* followed by repeated multiplication, it is known as micropropagation. It is a kind of vegetative propagation which is done under controlled laboratory/greenhouse conditions. Micropropagation generally produces plants which are "true to type". It is commonly used to multiply millions of plants from selected mother plants in horticultural and forestry. Commercial tissue culture is nearly synonymous with micropropagation.

## Applications of Tissue Culture Technology

Tissue culture technology is useful both in research and commercial propagation of plant material. In other words, tissue culture technique can be useful directly for the mass propagation or indirectly for varietal development. Following are some of the important applications in agriculture and horticulture

- Micropropagation of ornamental, fruit and forest tree species.
- Propagation of F1 hybrids of vegetables and other high value plants.
- Development of new varieties through somaclonal variation.
- Production of virus free plants through meristem culture.
- Production of wide crossed hybrids through embryo rescue.
- Synthetic seeds

- Cryo preservation of germplasm
- To regenerate plants after genetic modification (transgenics or GMOs)
- Production of secondary metabolites through cell culture.

## Micropropagation

Among various applications of tissue culture, the most widely used is the micropropagation. Traditionally, horticulture and forest plants have been improved and multiplied through conventional techniques of cuttings, grafting and air layering. However, all these techniques have some limitations in terms of number of propagules that can be produced from a single mother plant. In contrast, micro propagation allows round the year production of planting material from one identified mother tree, of course with some limitations. The most important one being that it needs to be done under controlled conditions with skilled manpower. The advantages and disadvantages of micro propagation are listed below

### *Advantages*

- Limited space needed
- Plants can be produced round the year
- "True to type" plants can be produced
- Disease (virus) free plants can be obtained
- Early flowering possible

### *Disadvantages*

- High capital, recurring costs and skill intensive
- Contamination control major challenge
- Variability possible some times (Eg. Multiple shoots, somatic embryogenesis)

Micropropagation or clonal propagation offers certain unique advantages as compared to seed raised planting material. The salient differences are given in the following Table 9.1.

**Table 9.1 Difference between seed raised and tissue cultured plants**

| Seed Raised Plants | Tissue Cultured / Clonal Plants |
|---|---|
| 1. Genetically variable gene | Uniform |
| Gene Complexes both for production (Short term economic yield) and future adaptability. Immediate fitness and long-term genetic flexibility. | Stress on short term economic yield at the expense of future adaptability immediate fitness only |
| 2. Low maintenance costs | High maintenance costs |
| Low inputs and moderate yield | High inputs and high yield |
| 3. Advisable under poly cultural forestry species-rich, various species serve different purpose/need | Advisable under monoculture conditions low genetic diversity |
| 4. Productivity low (per species basis) | High |
| High on cumulative basis | |
| 5. Advisable for conservation / revegetation / environmental forestry and wasteland development | Advisable for industrial forestry but selective for other types of forestry. |

## Development of Micropropagation Technology in India

India is one of the pioneering countries that started research on tissue culture including micropropagation. Protocols for large number of horticulture and forestry species have been standardized and scaled up. Some of the leading research Institutes which developed the tissue culture protocols include University of Delhi, National Chemical Laboratory, M.S. University of Baroda, National Botanical Research Institute (NBRI), ICAR

Research Institute, Baba Atomic Research Center in the public sector and Tata Energy Research Institute (TERI), Indo American Hybrid Seeds (IAHS) and A. V. Thomas and Co. in the private sector. Two pilot projects have been setup at TERI, New Delhi and NCL, Pune, which later became technology parks. In addition, DBT has also successfully implemented pilot projects for application of tissue culture in case of cardamom and vanilla. Due to the combined efforts of public and private sector, commercially viable protocols have become available and today several species like banana, sugarcane, teak, poplar and a large number of ornamentals like rose, gerbera, ficus and anthuriams are being produced commercially and marketed both in India and abroad.

## Problems in Technology Adoption

Although micropropagation technology has several advantages, it has still not picked up commercially in the country. India grows a variety of fruit and forest species, and a number of high yielding mother plants/varieties have been identified/developed by the scientists in agriculture and forest research institutes, which can be propagated through tissue culture, the scope is immense. However, there are a number of practical constraints in wider adoption of this technology, the most important being the high cost of energy in the country which forces the producers to price the products higher than the small and marginal farmers can afford. For a typical ornamental micro propagation project world wide, the cost break down works out to be 40 per cent for manpower, 20 per cent for energy, 10 per cent for materials and 30 per cent for sales and marketing, while in India the energy costs are as high as 40-50 percentage and therefore most companies are unable to invest on sales promotion, marketing or extension service. Other problems include, availability of alternative methods of propagation, which are cost competitive. Since most agricultural crops are season specific, there are peak demands in few months, which is difficult to schedule both for production and stocking. Poor infrastructure, transport and storage constraints also add to the problems in delivery of this technology at farmers level. These limitations have offset the advantages, the country enjoys in having skilled manpower. There is a growing world market for tissue-cultured ornamentals which India is unable to exploit due to the above constraints. Even small countries like Singapore and Thailand are ahead of India. Some of the leading firms in the private sector have not been able to sustain the high costs and gradually reduced their activities in tissue culture. Few export oriented units (EOUs) have started off well, but problems cropped up on the quality of the shipments in the receiving countries and in some cases infringement of IPRs on the mother plants. Very few are surviving now.

India has a vast internal market for fruit seedlings, hybrid vegetables and forest plants. There is a need to popularize the micro propagation technology, more importantly to achieve cost reduction, so that these plantings stock are within the reach of small and marginal farmers. A continued domestic market only can sustain this industry in the long term. Sugarcane, banana, papaya, ornamentals, medicinal plants and teak etc., offer good viable opportunities within the country due to the growing awareness on cultivation of improved planting material. Unlike other agro based units, micro propagation industries have to be less cost intensive, decentralized and address local needs by producing planting material for a given agro-climatic condition and set up good extension service in the target area.

## Some Common Perception on Tissue Cultured Plants and the Scientific Facts

Tissue culture technology has moved to the fields during the last decade. Banana, sugarcane, teak and ornamentals are some of the species routinely produced by public and private sector laboratories and marketed to the farmers. There are some apprehensions and misconceptions among farmers and extension workers on the merits and demerits of tissue cultured plants vis-à-vis seedlings or vegetatively propagated plants. The most important points are clarified in Table 9.2.

**Table 9.2 Perceptions and facts about tissue cultured plants**

| Perception | Fact |
|---|---|
| Tissue cultured plants are much superior to seedlings or vegetatively propagated plant material | The superiority depends on the mother plants used. Since most of the plants produced through tissue culture are already selected, based on performance, they might yield more as compared to local germplasm. The main advantage from tissue culture comes from uniformity in growth and yield which gives an overall economic advantage on area basis, but micropropagation per se does not result in superior planting stock. |
| Tissue cultured plants are disease/pest resistant | Plants do not become resistant merely by tissue culture. However, some species produced through meristem culture will be virus free alteast for few generations. Basically, tissue culture cleans the plant from inherent pathogens so that the plants will be vigorous for few generations. |
| Tissue culture plants do not survive in the field conditions | Not True. Properly handed tissue culture plants with good root system will have same survival rates as that of seedlings or vegetatively propagated plants. |
| Tissue culture timber plants like teak mature much earlier than the stumps | Not true. The maturity depends on the mother plant, the growth rate and interaction with the climate. There is no data to show that tissue cultured timber plants can be harvested earlier than normal plants. |
| Tissue cultured plants are fast growing as compared to stumps and seedlings | Rate of growth is the character of the mother plant. It does not increase due to tissue culture, though initial vigorous growth may be possible because of clean and healthy planting stock. |
| The root system in tissue cultured plants is not well developed and suitable for field survival | Tissue culture is a method of vegetative propagation. The roots are induced by hormone treatment of micro-shoots. Therefore, like in all other vegetative propagated plants (except on *in situ* grafted or budded plants), the root system will be adventitious. This type of root system has both merits and demerits depending on the life cycle of the plant and the harvest cycle. Tap root system, no doubt gives greater anchorage but even the adventitious root system in perennial trees eventually develops into a set of strong roots, which provide good support. |
| Tissue culture method of propagation causes variations in the plants produced | It is true in some cases. When the propagation is based on nodal explants or meristems, the chances of variation are extremely low. However, when other methods like somatic embryo genesis involving a callous phase is used, there are chances for either negative or positive variations. |

To summarize, the planting material of any species does not become superior merely because it is produced through micro propagation except when it is used to produce virus free plants. The quality of the planting material mainly depends on the mother plant. Micropropagation is one of the several methods available to the scientists and nursery men for vegetatively propagating elite plants. However, it is a powerful and unique technique where large number of propagules can be produced in a short time and space. The choice of method of propagation has to be made depending on the type of natural regeneration of given species, recalcitrance if any and the need for quality planting material at competitive cost.

## Bibliography

1. Biotechnology in Horticulture and Forestry, by Vibha Dhawan, Plenum Press, New York, 1989.
2. Dryland Agriculture in Andhra Pradesh: Needs Assessment and Options for Biotechnology – A report by Institute of Public Enterprise, Hyderabad, 1994.
3. Plant Biotechnology by S.S.Purohit, published by Agro Botanical Publishers (India), Bikaner, 1997-1998.
4. Status of Plant Biotechnology in India – An Assessment from the Small Farmer Perspective (A Study Sponsored by AP-NL Biotechnology Programme) by Naresh Sharma & P. S. Janaki Krishna, Institute of Public Enterprise, OU Campus, Hyderabad – 500 007, India, 1999).
5. Plant Tissue Culture: From Research to commercialization – A decade of support, Department of Biotechnology, Ministry of Science and Technology, Government of India, November 2000.

# 10

# Animal Vaccines and other Recent Advances

* M. Ramakrishna Rao

A live vaccine is the ideal vaccine because it replicates in the host, giving rise to wide range of immune responses simulating a natural infection. It tends to survive longer in the host and thus stimulation is prolonged. Another feature of live vaccine is the possibility of delivering it on to mucous membrane for induction of mucosal immunity, which is advantageous for animals that are vaccinated in the presence of maternal antibodies. However, the live vaccines may possess the risk of residual virulence for target host and some vaccines may revert to virulence and cause disease after vaccination, more so with single stranded RNA viruses. They always present the risk of contamination with unwanted organisms or other disease producing virulent organisms.

* Join Director (AH), Disease Investigation Wing. Veterinary Biological & Research Institute, Shantinagar, Hyderabad.

The killed vaccines can only meet the basic requirements of an ideal vaccine like induction of long lasting protection and harmless. The antibody titres after vaccination with killed vaccine may be higher than vaccination with live vaccine and consequently the antibodies to offspring through colostrums will also be high and so are passively protected better and longer. They are safe with respect to residual virulence and reversion to virulence. But the killed vaccines require repeated vaccinations to induce high level of protective immunity as they trigger only humoral response. The adjuvants used long with killed vaccines may cause severe local and systemic reactions and increase risk of hypersensitive reactions.

The advantages and disadvantages of any killed or live vaccines correspond to the disadvantages and advantages respectively. The choice of live or killed vaccine varies from one infectious disease to other and is dependent on molecular biology of microorganism, the pathogenesis and immunology of the disease and prevalence of the infection. As the pathogenesis and epidemiology of each disease varies, the role and efficacy of vaccination as a means of control also varies from one disease to other. Ultimately efficacy and safety are two main criteria to be considered for vaccine choice.

## Biotechnologically Designed Vaccines

The introduction of recombinant DNA and hybridoma technology has revolutionized the development of modern vaccines. These biotechnologically designed vaccines are alternatives to the conventional live, attenuated replicating vaccines and killed non-replicating vaccines.

The biotechnologically designed modern vaccines include recombinant DNA vaccines like synthetic peptide vaccines, deletion mutant live vaccines, reassortant live vaccines, vaccine vector live vaccines and anti-idiotype vaccines.

### Synthetic Peptide Vaccines

The amino acid sequence of the immunogenic fraction of the protein is the basis of synthetic peptide vaccine. The disadvantage of the synthetic peptide vaccine will be due to differences in configuration of synthetic peptide and the same peptide when it forms part of virus particle.

### Deletion Mutant Live Vaccines

By deleting the non-essential virulence genes from viruses or bacteria the deletion mutants are produced only from the organisms, which can multiply even on deletion of certain fraction of genes. These vaccines are safe and as immunogenic as conventional vaccines.

### Reassortant Live Vaccines

These vaccines are prepared from the viruses that have segmented genomes. By simultaneously infecting cell cultures with two different viruses through a process of recombination reassorted strains may emerge. But it is important that the desired reassorted strain should be identified before using it as vaccine.

### Vaccine Vectored Live Vaccines

In recombinant DNA vaccines the gene coding for the immunologically important protein of the candidate strain will be identified and inserted into other biological systems, which grow in culture and yield high quantities of immunogenic protein. Vaccine and Poxviruses are usually as expression systems. *E.coli* and Saccharomyces cervicisiae are also being used as vectors.

### Anti-Idiotype Vaccines

Anti-idiotype vaccines are based on the principle that the antigen-combining site (idiotype) of an antibody can function as antigen and therefore elicits antibodies directed against the idiotype, thus termed anti-idiotype antibodies. The anti-idiotypes can also mimic non-protein structures such as carbohydrates, which cannot be produced by gene cloning.

Biotechnologically designed vaccines are classified into three categories. This classification is based on product, biological properties and on the safety concerns they present.

Category – I vaccines include non-viable or killed products that pose no risk to the environment and present no new or unusual safety concerns.

Category – II vaccines include vaccines which have microorganisms modified by addition or deletion of one or more genes. Such genetic modification should not cause deterioration in the safety character rustics of the organism.

Category –III vaccines include live vectors to carry recombinant – derived foreign gene that code for immunizing antigens.

## Advantages of Biotechnologically Designed Vaccines

They can be designed for organisms that do not grow or show limited growth in cultures. They are physically stable and heat resistant. Incomplete inactivation, which occurs in conventional vaccines, may be avoided. Contamination with extraneous agents is less. They have low amounts of tissue culture components and other immunologically active components. They can be produced against parasitic diseases also, for which conventional methods cannot be used. Large-scale culture of hazardous agents can be avoided like in Rabies with biotechnologically designed vaccines. With vaccine vectored viruses the protection levels induced will be high because the vector multiplies and it can be used as multivalent vaccine vector. Such vaccines are cheap and easy to produce and easy to administer.

# 11

# GM Technology in Agriculture

[1] N. P. Sarma and [2] S. M. Balachandran

Genetic modification of plants has been the basis for betterment of crops in modern agriculture. Since prehistoric times, farmers have practiced seed selection that resulted in better plants and crops. Hybridization followed by selection has led to the evolution of a wide range of crops and varieties. The power of these practices was greatly enhanced in the 20th century with the application of the science of genetics leading to modern hybrid varieties of food crops like maize, sorghum, pearl millet and high yielding inbred varieties in self-pollinated crops like rice and wheat. Despite the spectacular success in agricultural growth and food production largely ushered by Green Revolution technologies, the sustainability of such production advance is threatened due to population increase, depletion

[1] Formerly Principal Scientist, Directorate of Rice Research, Hyderabad – 500 030.

[2] Scientist, Directorate of Rice Research, Hyderabad – 500 030.

of natural resources – land and water and destabilizing factors such as biotic and abiotic stresses. Amidst these challenges the only way to expand production is to develop and deploy technologies that enhance output per unit of the input. The recent developments in new biology collectively known as Biotechnology promise to not only enhance productivity further, but also to address a wider range of constraints to production both in high production environment and also in grey and fragile areas.

Plant Biotechnology which includes genetic engineering through transfer of genes from unrelated plants and microorganisms (GM technology) has had profound impact in Agriculture and one of its most important applications has been the development of transgenic crops or more popularly known as genetically modified (GM) crops. Many millions of hectares of GM crop varieties of soybean, cotton, canola, maize, are being grown annually. As on 2003, the global area under transgenic crops is about 67.7 million hectares, grown in 18 countries including USA, Argentina, Canada, Brazil, China and South Africa etc. In the developing world, GM crops have had largest impact in Argentina, Brazil, China and South Africa. India has recently approved the release of first GM crop of cotton for cultivation in farmers' field.

Since the development of transgenic crops in the 1980s and their commercialization sometime in mid-1990s, the focus was on market-oriented traits in commercially important crops like soybean, cotton and canola. GM technologies have benefited farmers by reducing the cost on pesticides and on chemical weed control. No GM varieties of a major food grain crops like rice or wheat are currently grown anywhere. Nonetheless, the GM technology raises much hope to ensure sustainable food production in addition to combating malnutrition.

## GM Technology

Genetic engineering is one of the components of biotechnology, which enables genes that carry instructions for a particular trait to be isolated and moved from one organism to another. A transgenic plant contains a gene or genes isolated from different source that has been artificially inserted instead of the plant acquiring them through pollination. The inserted gene, known as transgenic, may come from another unrelated plant or from any taxon. Plant with transgenic(s) is often called 'genetically modified plant' (GM plant) or 'genetically engineered plant' (GEP). The uniqueness of GM technology lies in the fact that gene transfers are not limited by reproductive barriers.

## The GM technology Involves the following steps:

(i) Generation of DNA fragments, cloning into a 'vector' or vehicle and creation of a suitable gene construct.

(ii) Transfer of the gene construct into plant cells/tissues through a process called 'genetic transformation'.

(iii) Selection of transformed cell lines or seedlings using a suitable marker system and regeneration of fertile plants from the transformed cells.

(iv) Molecular analysis of transformed plants for stable integration, expression and inheritance of transgenic(s).

## Methods of Gene Transfer into Plants

With the development of recombinant DNA technology or genetic engineering, we have now means to introduce 'novel' genes from a variety of sources into target crops. Methods of gene transfer include (i) use of plant pathogenic bacterium - Agrobacterium tumefaciens as a vector, (ii) DNA coated metal micro projectiles and (iii) direct uptake of DNA by protoplasts of plant cells. The main advantage of rDNA technology is that unlike hybridization it facilitates transfer of specific well-characterized genes from source organism to the target crop variety. Tremendous progress has been made in application of GM technology in Agriculture in the past two decades.

## Genetic Engineering of Crop Plants for Herbicide Resistance

Engineering herbicide resistance of crop plants has been the most successful area of GM technology applications. Tolerance/resistance to herbicides could be achieved either through altering the level and sensitivity of the target enzyme to the herbicide or introducing the gene that can detoxify the herbicide in the plant. Using such strategies herbicide resistance in different crops, particularly soybean, corn, cotton, rapeseed, mustard, rice, maize, sugar beet and alfalfa has been incorporated (Rai and Prasanna, 2000). The specific herbicides for which tolerance has been engineered are phosphinothricin (Basta$^{TM}$, Herbiace$^{TM}$), glyphosate (Roundup$^{TM}$), sulfonylurea (Glean$^{TM}$) Bromoxynil (Buctrill$^{TM}$) and 2,4-dichlrophenoxy acetic acid (2,4-D).

## Engineering Crop Varieties for Biotic Stresses

Biotic stresses caused by insects, fungal, bacterial and viral pathogens inflict heavy losses to crop productivity. Deployment of host plant resistance and chemical control through pesticide application are the ways to combat the losses. Adequate gene sources imparting resistance to certain pests and diseases may not always be available in the germplasm to breed for host plant resistance. Genetic engineering has a specific role in this regard when there is no satisfactory level of resistance either in the primary or secondary gene pools of crops for exploitation through plant breeding.

## Insect/Pest Resistance

Development of insect resistant transgenic crops is one of the demonstrated successes of genetic engineering in agriculture. Transgenic plants carrying the insecticide protein gene from Bacillus thuringiensis (Bt) has been by far the most successful example. Bt is a soil bacterium, which synthesizes crystalline protein during sporulation. These crystal proteins are solubilized in the highly alkaline midgut of the larvae to release protoxin that interferes with the osmotic equilibrium and cell lysis ultimately leading to insect paralysis and death. The first transgenic tobacco plants with Bt were produced in 1987, and subsequently several crop plants including cotton, corn, potato, rice, tomato, eggplant, canola, soybean etc., have been developed with Bt genes. As of now, more than 30 plant species that also include peanut, alfalfa, apple, white clover, broccoli, grapes, pear and sugarcane have been transformed with Bt cry genes (Ranjekar et al. 2003). Aside Bt genes, vegetative insecticidal proteins (VIP), proteinase inhibitors, plant lectins, á-amylase inhibitors and insect chitinase genes have also been tried in various crop plants for production of transgenics against insect pests.

## Disease resistance

Considerable yield loss in most of the agri-horticultural crops is accounted due to diseases caused by fungi, bacteria and viruses. In India, a recent survey by Grover and Gowthaman (2003) reports that fungal diseases are rated as the major or at least second most important stress that contributes to yield losses in major cereals, pulses, and oil seed crops. Breeding for resistant varieties and use of agrochemicals are some of the strategies being followed in controlling pathogens. However, frequent break down of resistance in varieties, evolution of new races of pathogens and residual effect of agrochemicals still pose serious challenge to crop protection. Transgenic lines that express antifungal, antibacterial proteins have been successfully produced in several crop plants. Among the genes that encode defense response proteins, pathogenesis-related (PR) proteins, plant ribosome-inactivating proteins (RIPs), lipid transfer proteins (LTPs), polygalacturonase inhibitor proteins (PGIPs), antiviral proteins etc., have been widely used for developing transgenic plants of many species. Following the first report of introduction of bean chitinase gene into tobacco and Brassica against Rhizoctonia solani by Broglie et al. (1991), there have been several reports on transgenics developed with different PR proteins. The important crops where PR genes were introduced to develop transgenics against different fungal pathogens include, rice, tobacco, Brassica, potato, tomato and wheat. Some secondary metabolites such as phytoalexins are believed to have a role in plant defense. Transgenic plants with phytoalexins developed in rice, tomato, barley and wheat were shown to increase resistance to different fungi. Resistance genes involved R-Avr interaction has been isolated from many crops and fungus-resistant transgenics are being produced by incorporating the R genes in susceptible plants within a genus or a family or even outside the family (Grover and Gowthaman, 2003). Virus resistant transgenics have been developed in a number of crops by introducing either viral coat protein (CP) or replicase gene encoding sequences. A large number of transgenics in cereals, fruit crops, vegetable crops and legumes are reported to be developed through coat protein mediated resistance (Dasgupta et al. 2003).

## Tolerance to Abiotic Stress

Abiotic stresses, particularly drought, salinity and low temperature are the major constraints for increasing crop production. The phenotypic and physiological status of the plant under stress is determined by accumulation of sugars and other solutes and osmotic adjustments. The complex genetic mechanisms of these stresses are governed by polygenes. Several stress related genes are now cloned and transferred to different crop plants for obtaining transgenics resistant/tolerant to drought, salinity and cold. Transcription factor genes such as abscisic acid induced protein (abi3), dehydration response element (dreb1A), soybean cold inducible factor (scof1); signal transduction component genes- superoxide dismutase (sod), serine acetyl transferase (sat), glutathione reductase (gr); fatty acid metabolism genes - glycerol 3, phosphate acetyltransferase (gpat); heat shock genes (hsp101); osmolyte biosynthesis genes-choline dehydrogenase (betA), Choline oxidase A (codA), Trehalose-6-phosphate synthetase gene

(TPS1), pyrroline carboxylate synthetase (p5cs), Pyruvate decarboxylase (pdc1), have been cloned and used for developing transgenic plants. Transgenics resistant/tolerant to drought, salinity and cold are already developed in Arabidopsis, tobacco, alfalfa, tomato, rice etc.

## Nutritional Enhancement through Genetic Engineering

Plants provide almost all the essential vitamins, amino acids, minerals and several other micronutrients essential for human health. Unfortunately, for majority of poor people in the developing countries foods with all the essential nutrients are inaccessible. Their dependency only on staple food like rice, wheat, which contains low levels of micronutrients, leads to malnutrition among the poor people. Improving the nutritional quality of crop plants is set to become a growing area of application for GM plant breeding. Micronutrient enriched rice with iron (Goto et al. 1999), β-carotene (Ye et al. 2000) and lysine by genetic engineering is now a reality. A well-known example is the development of a rice variety with an enriched vitamin A and iron content. It was developed by an international working group to contribute to a better vitamin A supply for millions of people today suffering from deficiencies leading to different clinical dysfunctions mainly in children and pregnant women in the developing countries. The so-called "Golden Rice" is being tested for bioavailability of vitamin A and allerginicity etc. Improvement of nutritional status of some of our major crops is one of the thrust areas of transgenic research in India. To begin with, work on three major crops - rice, wheat, and potato has already begun. High-yielding rice varieties grown in India will be either transformed with constructs that would code for high β-carotene or by back crossing these varieties to "Golden rice" transgenics already developed in Switzerland and Germany. Improving iron content in rice grains will be another thrust area. Nutritionally enhanced transgenic Indian varieties of potato carrying the ama1 gene are already under advanced field trials. If found suitable, these will be released for cultivation (Sharma et al. 2003).

## Global Status of Transgenic Crops

The rapid adoption of transgenic crops since the first GM crops were introduced in 1996 bears testimony to the growing demand for the GM crops. Despite the controversies surrounding the GM technology, adoption rates for transgenic crops are unprecedented and are the highest for any new technology by agricultural industry (James, 2002).

Since the introduction of transgenic crops for commercial cultivation in 1996 with 1.7 million hectares, the global GM crop area continued to grow for the sixth consecutive year at a sustained rate of more than 10 per cent. The estimated global area of transgenic crop or GM crops for 2003 is 67.7 million hectares (Fig. 11.1). Almost one-third (30%) of the global GM crop area of 67.7 million ha. was grown in 11 developing countries. India, the largest cotton growing country in the world, commercialized Bt cotton for the first time in 2002.

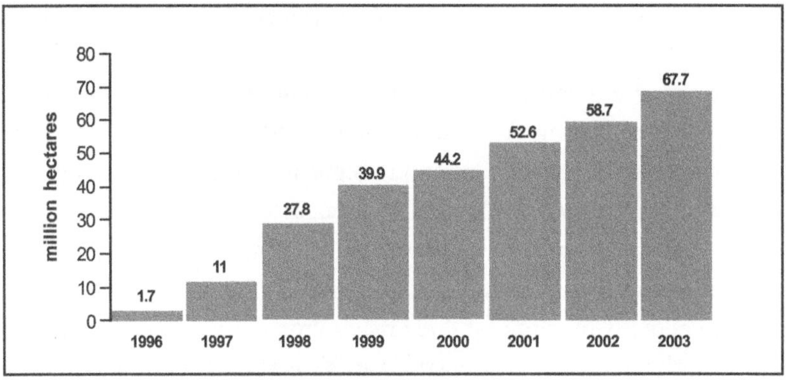


**Fig. 11.1** Global Area of Transgenic Crops (1996 - 2003).

The four principal countries that grew 99 percent of the global transgenic crop include USA with 39.0 million ha. (66% of global total), followed by Argentina with 13.5 million ha. (23%), Canada 3.5 million ha. (6%), and China 2.1 million ha. (4%) (Table 11.1). China had the highest year-on-year percentage growth with a 40 percent in its Bt cotton area, which occupied more than half (51%) of the national cotton area. During the six-year period 1996 to 2002, herbicide tolerance has consistently been the dominant trait (75%), with Bt insect resistance second (17%) and stacked genes third (8%). Among the GM crops, herbicide tolerance soybean, Bt cotton, canola, and Bt maize were the major GM crop, grown worldwide during 2002. There is cautious optimism that global area and the number of farmers planting GM crops will continue to increase in 2003 (James, 2002).

**Table 11.1 Global area of transgenic crops, (2002) by country-wise (million hectares)**

| Country | Million hectares | % |
|---|---|---|
| U.S.A | 42.8 | 63 |
| Argentina | 13.9 | 21 |
| Canada | 4.4 | 6 |
| Brazil | 3.0 | 4 |
| China | 2.8 | 4 |
| South Africa | 0.4 | 1 |
| Australia | 0.1 | < 1 |
| India | < 0.1 | < 1 |
| Romania | < 0.1 | < 1 |
| Spain | < 0.1 | < 1 |
| Uruguay | < 0.1 | < 1 |
| Mexico | < 0.1 | < 1 |
| Bulgaria | < 0.1 | < 1 |
| Indonesia | < 0.1 | < 1 |
| Colombia | < 0.1 | < 1 |
| Honduras | < 0.1 | < 1 |
| Germany | < 0.1 | < 1 |
| **Total** | **67.7** | **100** |

**Source :** James, 2003.

Though majority of GM crops are grown in the developed countries, the adoption among the developing nations are limited to a few countries until now. However, the trend among the developing countries in the last few years shows that research on GM crop development and commercialization is slowly picking up. Particularly in many Asian countries GM crops are at different stages of development and countries like China, India and Indonesia have already entered into commercialization of GM crops (Table. 11.2).

**Table 11.2 Asian Scenario of GM Crops**

| Country | In Development | Field Trials Cultivation | Commercial |
|---|---|---|---|
| China | At least 20 crops including Cotton, corn, rice, potato, poplar, tobacco, groundnut, cassava, Chinese cabbage, sweet potato | Cotton, tomato, Sweet pepper, petunia | Cotton, tomato, sweet pepper, petunia |
| India | Rice, cotton, tobacco, potato, tomato, eggplant, chickpea, cauliflower, cabbage, papaya, maize, groundnut, cassava, coffee, muskmelon. | Cotton, rice, tobacco, corn, potato, mustard | Bt cotton |
| Indonesia | Not available | Cotton, potato, corn, soybean | Bt cotton |
| Japan | Rice, tomato, melons, broccoli, cucumber. | None | None |
| Malaysia | Rubber, oil palm, chillies, papaya, orchids. | None | |
| Philippines | Rice, banana, papaya, mango, coconut. | Bt corn | None |
| South Korea | Rice, leafy vegetables. | None | None |
| Taiwan | Rice, brassicas, tomato, peppers, mung bean, eggplant, onion, garlic. | None | None |
| Thailand | Rice, cotton, tomato, papaya, pepper. | Bt cotton, Papaya | None |
| Vietnam | Rice, papaya, potato, tomato, sugarcane. | None | None |

## Indian scenario of GM crops

Crop biotechnology research in India is mostly done by government and public sector institutions under different department/ministries. The Department of Biotechnology (DBT), Indian Council of Agricultural Research (ICAR) and Council of Scientific and Industrial Research (CSIR) are the major organizations that support biotechnology programmes. The DBT supports various activities of biotechnology in the country and a major portion of its budget – Rs. 2897 lakhs was allotted to Crop Biotechnology research alone during the IX$^{th}$ Plan period. These organizations have been supporting research projects on development and evaluation of transgenics with genes of agronomic importance. To bring researchers with expertise in different areas of biology together to address defined research goals, a number of multi-institutional projects have been launched (Sharma et al.2003). Considerable progress has been made in developing transgenic crops resistant to biotic stresses including insect resistant, resistant to diseases caused by fungi, bacteria and viruses. Major initiatives on research, development and testing of GM crops in both public and private sector institutes are listed in Table 11.3 and Table 11.4.

While the GM product development is the major goal of many research institutions across the country, the Government of India also acknowledges the importance of evaluation of GM crops for health and environmental safety. To address the general concerns of public on genetically modified crops, the DBT in collaboration with other government departments has developed a three tier regulatory system for biosafety evaluation of GM products. It comprises the following :

- Institutional Biosafety Committee (IBSC), set up at each institution for monitoring institute level research in genetically modified organisms.

**Table 11.3 Transgenic research and application in major Indian Institutes #**

| Institute | Transgenic Crop | Gene(s) inserted | Trait incorporated |
|---|---|---|---|
| Central Institute for Cotton Research, Nagpur. | Cotton | *Bt. cry* gene(s) | Resistance to Lepidopteran pests |
| Central Potato Research Institute, Simla. | Potato | *Bt. Cry1A(b)* | Resistance to Lepidopteran pests |
| Central Tobacco Research Institute, Rajahmundry | Tobacco | *Bt. cry1A(b), cry1A(c)* | Resistant to *Helicoverpa armigera* and *Spodotera litura* |
| Delhi University South Campus, New Delhi | Mustard/ rapeseed | *bar, barnase,* barstar | Herbicide resistance, Transgenic male sterile & restorer lines for hybrid seed production |
| | Rice | *Pyruvate decarboxylase & alcohol dehydrogenase* | Tolerant to flooding |
| | Rice | *codA, cor47,* | Resistance against biotic and abiotic stresses |
| | Tomato | *Ctx-B and Tcp* antigens of *Vibrio cholerae* | Edible vaccine development |
| | Brinjal | Chitinase, glucanase and thaumatin encoding genes | Resistance to diseases |
| | Wheat | *bar, HVA I, PIN2* | Resistance against biotic and abiotic stresses |

Table. 11.3 *Contd....*

| Institute | Transgenic Crop | Gene(s) inserted | Trait incorporated |
|---|---|---|---|
| Directorate of Rice Research, Hyderabad. | Rice | *Xa-21, cry1A(b)* | Resistance to yellow stem borer and bacterial blight disease |
| Indian Agricultural Research Institute, New Delhi. | Brinjal | *Bt cry1A(b)* | Resistance to Lepidopteran pests |
| | Tomato | *Bt cry1A(b)* | Resistant to Lepidopteran pests |
| | Cauliflower | *Bt cry1A(b)* | Resistant to *Plutella scylostella* |
| | Cabbage | *Bt cry1A(b)* | Resistant to *P.scylostella* |
| | Rice | *Bt cry1A(b), Cry1Ac* & Fusion *Bt* | Resistance to yellow stem borer |
| | Mustard/ rapeseed | *Arabidopsis* annexin gene | Stress-resistance/ tolerance |
| | Mustard/ rapeseed | *Choline dehydrogenase* | Abiotic stress-tolerance |
| | Potato | ACC synthase | Delayed fruit ripening |
| | Tomato | -do- | |
| | Banana | -do- | |
| | Tobacco *Brassica* | Chitinase, glucanase and *RIP* | Resistance to fungal pathogen |
| | Pigeonpea and *lectin* genes | *Protease* inhibitor | Resistance to bollworms and aphids. |

Table. 11.3 ***Contd....***

| Institute | Transgenic Crop | Gene(s) inserted | Trait incorporated |
|---|---|---|---|
| International Centre for Genetic Engineering and Biotechnology, New Delhi. | Tobacco | *Btcrylla5* | Resistance to *Spodoptera litura*. |
| | Cotton | *Bt* | Resistance to boll worm |
| International Crop Research Institute for Semi-Arid Tropics, Hyderabad | Chickpea | *PGIP* | Resistance to fungal pathogens. |
| Indian Institute of Horticultural Research, Bangalore | Tomato Chitinase and glucanase | *Leafcurl virus* sequence, | Resistance to leaf curl virus & fungal diseases |
| | Coat protein gene | | Resistance to Citrus triesteza virus |
| Jawaharlal Nehru University, New Delhi. | Potato | *Ama-l* | Nutritionally enriched potatoes |
| | Tomato | *OXDC* | Resistant to fungal infection |
| Madurai Kamaraj University, Madurai. | Black gram | Coat protein and replicase genes | Resistance to yellow mosaic virus |
| | Black gram | *Dianthin* & *barnase* Tolerance *bar* gene | Insect resistance and herbicide |
| | Rice | *Chitinase, osmotin* | Resistance to sheath blight fungus and *P-1,3-glucanase* |

**Table. 11.3** ***Contd....***

| Institute | Transgenic Crop | Gene(s) inserted | Trait incorporated |
|---|---|---|---|
| National Botanical Research Institute, Lucknow. | Cotton | *Cr1E and Cry1C* with terminal altered at C-end | Resistance to *Spodopoterlitura* and *Heliothis armigera* |
| Punjab Agricultural University, Ludhiana. | Rice | *Cry1Ab, Cry1Ac* | Resistance against yellow stem borer |
| Tata Energy Research Institute, New Delhi. | Mustard | *Ssu-maize Psy* and *Ssu-tpCrtl* gene | Oil with high levels of β-carotene |
| Tamil Nadu Agricultural University, Coimbatore | Rice | *chitinase* | Resistance to sheath blight (*Rhizoctonia solani*) |
| Osmania University, Hyderabad. | Rice | *gna* (lectin) | Resistance to Brown Planthopper |

#Adopted from Sharma et al. 2003.

**Table 11.4 Transgenic research and development activities in Indian private sector#**

| Name of the Institute/Company | Plants/Crops | Transgene(s) inserted | Trait (s) for evaluation |
|---|---|---|---|
| Ankur Seeds Ltd, Nagpur. | Cotton* | *Cry1A (c)* | Resistance to lepidopteran pests |
| Hybrid Rice International, Gurgaon. | Rice | *Cry1A(b), cry9C and bar* genes | Resistant to lepidopteran pests and herbicide |
| Indo-American Hybrid Seeds, | Tomato | *Alfalfa glucanase Tomato leafcurl Virus Genes* | Resistance to lepidopteran pests & Leaf curl virus |
| MAHYCO, Mumbai. | Cotton* | *Cry1A(c)* | Resistance to lepidopteran pests |
| | Cotton* | CryXgene | Resistant to lepidopteran pests |
| | Rice | Cry1Ac, Xz21 & GNA gene | Resistance to yellow stem borer and sucking pests |
| | Mustard | CP4 EPSPS | Resistance/tolerance to herbicide |
| MAHYCO Research Foundation, Hyderabad. | Rice | *Xa21* gene | Resistance to bacterial blight |

Table. 11.4 *Contd....*

| Name of the Institute/Company | Plants/Crops | Transgene(s) inserted | Trait (s) for evaluation |
|---|---|---|---|
| ProAgro PGS (India) Ltd, Gurgaon | *Brassica* mustard | *Bar, barnase, barstar* | To develop superior hybrid cultivars<br>To develop superior hybrid cultivars |
| | Tomato* | *Cry1A(b)* | Resistance to lepidopteran pests |
| | Brinjal* | *Cry1 A (b)* | Resistance to lepidopteran pests |
| | Cauliflower* | *Cry1Hlcry9C* | Resistance to lepidopteran pests |
| | Cauliflower* | *Bar, barnase, barstar* | To develop superior hybrid cultivars |
| | Cabbage* | *Cry1H/cry9c* | Resistance to lepidopteran pests |
| Syngenta India Ltd, Pune | Cotton* | *Vip-3 gene* | Resistance to lepidopteran pests |
| | Maize* | *Cry1A (b)* | Resistance to lepidopteran pests |

*Transgenics developed elsewhere are being backcrossed to inbred lines.

#Source: Sharma et al. 2003.

- Review Committee on Genetic Manipulation (RCGM), functioning at DBT to monitor ongoing research activities in GMOs. A Monitoring and Evaluation Committee (MEC), comprising agricultural scientists, was constituted in 1998 by RCGM to monitor and supervise field trials permitted by the government.
- Genetic Engineering Approval Committee (GEAC) in the Ministry of Environment and Forestry with the involvement of DBT and ICAR has been set-up to authorize large scale trials and environmental release of genetically modified organisms (Ghosh and Ramanaiah, 2000).

## Biosafety concerns

The idea of transferring a gene from an unrelated organism using recombinant DNA techniques inherently entails greater risks than traditional cross breeding raised many potential concerns. The weight of scientific evidence though leads to the conclusion that there is nothing to substantiate this view, the success and benefit to the society at large from GM technology would depend on the public acceptance of GM foods. It is therefore, important that all questions related to perceived environmental risks from GMOs as well as safety of GM foods are properly addressed through satisfactory science based answers. Biotechnology has had its share of detractors and activists who claim GMO's are dangerous to health and environment. The myths that have emerged with the advent of this technology are:

- Emergence of new species that turn out to be super weeds
- Evolution of pests that resist chemicals and other forms of protection
- Harm to ecosystem through destruction of non-target species
- Threat to Biodiversity

- Resistance to antibiotics in humans through GM foods
- Benefit to producers but not to consumers
- Private sector monopoly of GM Technology

First of all, it should be understood that no technology is with 'zero risk'. Secondly, the greatest threat to bio-diversity is in expansion of agriculture land and any technology that can help stop this and minimize the chemical inputs will probably be that of by GM. Further, a large proportion of developing world agriculture is in the hands of small-scale farmers and whose interest must be safeguarded. If the food needs of expanding population are to be met, sustainable practices in Agriculture must be introduced without attendant damage to environment or natural resources. The potential of transgenic crops as a means to enhance agricultural productivity has been accepted by several countries as evident from their extent of adoption worldwide barring those of European Union. Nevertheless, because of the revolutionary nature of creating novel plants the likely risk and uncertainty may be created as a result of new and unrelated DNA sequences inserted into target varieties. In addition, it is perceived that the use of antibiotic resistance gene as selectable marker in the production of transgenics pose serious implications to human health especially when products from genetically modified plants enter the food chain. These and many more doubts plague the mind of public at large. Therefore, if the potential of GM technology is to be realized, it has to be done in a responsible manner with proper testing and regulatory mechanisms. There is a pressing need for scientists, policy makers, NGO's, extension workers and progressive farmers to come together on a common forum to discuss some of the concerns about environmental safety, socioeconomic effects, ethical and also regulatory issues. Fortunately, India is one among the few countries to adopt biosafety guidelines, with the involvement of concerned ministries and agencies. Several public sector institutions and private companies are already involved in evaluation and development of strategies for deployment of transgenics in the country. In the year's ahead GM technology has a greater stake in increasing food production, improve the efficiency of food production and reduce the environmental impact of agriculture.

To summarize, in the field of agriculture several crop plants have been genetically modified to have resistance to insects (cotton, maize, potato, melons and vegetables) and others to have resistance to herbicides (cotton, maize, oilseed, rape seed, wheat and soybean) and yet others to viruses and fungal pathogens (rice and potato), quality of foods has been improved (sunflower and canola), essential vitamins and amino acids added (golden rice, iron rich rice and lysine rich maize), shelf life of fruits enhanced (tomato and straw berry). In addition to pest and herbicide resistance, other traits that confer drought and salt tolerance are being added to crop plants through this technology. Future advances offer the promise of an impressive array of new and useful products that will enhance yield, quality, provide better nutrition, deliver needed vaccines (edible vaccines), produce more desirable fats and oil, extended shelf life of fruits and vegetables. It should be emphasized that the advantage of the transgenic crops solely lies on the farmers side. The first waves of transgenics are directed in favour of producers and farmers. The consumer hardly derived any direct benefit which is why public apathy for the biotechnology. With the shift in beneficiary profile, the situation may soon change when next generation transgenic crops enter the market. The successful development of the nutritionally rich rice especially "golden rice" and iron-fortified rice would be crucial. It is expected that acceptability of these value added rice would ultimately pave way for easy introduction of other transgenic products in the market.

***References***

Dasgupta, I., Malathi, V.G., and S.K. Mukherjee. (2003). Genetic engineering for virus resistance. *Current Science*, Vol. 84 (3): 341-354.

Ghosh, P.K., and T.V. Ramanaiah. (2000). Indian rules, regulations and procedures for handling transgenic plants. *J. Sci. Ind. Res.*, 59: 114-120.

Goto, F., Yoshihara, T., Shigemoto, N., Toki, S., and F. Takaiwa. (1999). Iron fortification of rice seed by the soybean ferritin gene. *Nature Biotechnology*. 17: 282-286.

Grover, A., and R. Gowthaman. (2003). Strategies for development of fungus-resistant transgenic plants. *Current Science*, Vol. 84 (3): 330-340.

James, C. (2003). Global status of commercialized transgenic crops: 2003. ISAAA Briefs No. 30 ISAAA: Ithaca, NY.

Rai, M., and B.M. Prasanna. (2000). *Transgenics in Agriculture*. Indian Council of Agricultural Research, New Delhi, India.

Ranjekar, P.K., Patankar, A., Gupta, V., Bhatnagar, R., Bentur, J.S., and P. Ananda Kumar. (2003). Genetic engineering of crop plants for insect resistance. *Current Science*, Vol. 84 (3): 321-329.

Sharma, M., Charak, K.S., and T.V. Ramanaiah. (2003). Agricultural biotechnology research in India: status and policies. *Current Science*, Vol. 84 (3): 297-302.

Ye, X., Al-Babili, S., Kloti, A., Zhang, J., Lucca, P., Beyer, P., and I. Potrykus. (2000). Engineering the provitamin A (â-carotene) biosynthetic pathway into (carotenoid free) rice endosperm. *Science*, 287: 303-305.

# UNIT - III

## Issues in Biotechnology Applications

# 12

# Socio-Economic and Ethical Implications of Various Agricultural Biotechnology Interventions

* K. P. C. Rao

Biotechnology is the application of biological methods or processes to produce useful products. The ancient Egyptians were credited with the invention of fermenting beer by using yeast. Indians have been making curds by using biological cultures for centuries. Several conventional biotechnology processes and approaches are in use in India. Some of them are mentioned below:

1. Fermentation or conversion of substrates into desired products by biological processes.
2. Downstream processing for recovery of metabolites.
3. Use of microbes or enzymes for producing value added products.
4. Sera, vaccines and diagnostics produced by conventional methods.

* Principal Scientist (Agl. Econ), ICRISAT, Patancheru, Andhra Pradesh.

5. Reproduction, artificial insemination and embryo transfer technology for animal breeding.
6. Methods for fish spawning induction.
7. Plant cell or tissue culture.
8. Biofertilizers.
9. Biopesticides.
10. Shelf life improvement.
11. Industrial waste treatment.
12. Plant breeding based on marker aided selection.
13. Vermi culture

Farmers have been selecting and sowing the seeds from plants with beneficial characters, such as higher yield, better nutrition and resistance to diseases and pests etc.

Unknowingly, they have been modifying the genetic make up of plants and animals, albeit slowly. The power of these practices was enhanced dramatically in the twentieth century by breakthroughs in basic science of genetics, leading eventually to modern hybrid seed varieties for important food crops such as maize and by mid century, to high yielding "Green Revolution "seed varieties for wheat and rice.

In 1973, scientists began engineering recombinations of DNA molecules by moving specific genes carrying desired traits from a source organism into the DNA of a living target organism. This genetic engineering technique, which is commonly referred as genetic modification (GM), seemed to promise not only greater range and speed for genetic modification processes but also greater control over the outcomes.

However, it took more than two decades to develop and commercialize GM crops during the mid-nineties of twentieth century. Due to the 'permissive' policies followed by United States, Canada and Argentina, the GM varieties of corn, cotton, soybean etc., spread rapidly in these countries. But in Europe, where 'green' parties are strong and where a ' mad cow disease' crisis in 1996 sensitized the media to food safety issues, the GM crop revolution encountered strong social resistance. European Governments are following 'precautionary' policies and have imposed separate labeling requirements on GM foods since 1997. These divergent policies toward GM technologies in the developed countries have now created a complicated problem of policy choice in the developing world. Caught between the bulls and bears of GM crops, scientists in the developing countries are engaged in the task of maximizing the benefits from GM crops while keeping the risks down to the acceptable levels.

The Green Revolution technologies were universally accepted without much protest or opposition. The plant breeding methods that were used to evolve high yielding varieties and hybrids were considered 'natural' as there was no transfer of genes across species. But the genetic engineering techniques employed in GM crops also involve transfer of genes between species and sometimes, even from plants to animals and the *vice versa.* Such a transfer of genes between unrelated species might appear ' unnatural' and 'immoral' to the conditioned psyche of people in the developing countries. Besides the perceived risks of GM crops affecting the health of humans and animals, these 'moral' or 'ethical' concerns about some techniques of modern biotechnology such as cloning and genetic engineering are contributing to the mounting opposition to GM crops and modern biotechnology. But the eventual acceptance of these technologies would depend upon the objective merits and demerits of these technologies rather than on the subjective feelings, which may subside as more and more such breakthroughs occur. But, at the moment, the socio-economic and ethical concerns about the new agricultural biotechnology interventions may be limiting their acceptance by the producers and consumers.

## I. Cost-Benefit Analysis of Biotechnologies

There are not many studies that attempted a systematic economic analysis of different biotechnology innovations. However some rough estimates and *ex ante* economic analyses are available. An attempt is made to collect and present these estimates to give an indication of their economic viability.

### (A) Vermicompost

Vermi composting is an effective process of recycling farm residues by using earthworms to increase humus content in the soil. Earthworms can convert about 1000 tonnes of moist organic wastes into 300 tonnes of rich dry vermi compost. In 45 to 60 days one Kg of earthworms can produce roughly 10 kg of vermi-castings. The economics of a commercial vermi compost unit are summarized in Table 12.1.

**Table 12.1 Economics of a Vermiculture Unit in ½ acre site**

| Non-Recurring Expenses | Rs. |
|---|---|
| 1. Construction cost of sheds (180 x 6 x 5ft) | 60,000 |
| 2. Cost of sprinkler system – 1 HP motor and water tank of 250 litres capacity | 1,60,000 |
| 3. Borewell, pipes, Pump + 5 HP motor | 1,60,000 |
| 4. Electrically operated cutting machine including motor | 15,000 |
| 5. Cost of planks, spades, Bamboos etc | 10,000 |
| 6. Compost storage shed of 1500 s. ft | 60,000 |
| 7. Weighing machine, motor cycle, Telephone etc | 30,000 |
| **Total** | **4,95,000** |

Table 12.1 *Contd...*

| Recurring Expenses | Rs. |
|---|---|
| 1. Salary of a manager and a salesman @ Rs 5000/pm | 60, 000 |
| 2. 4 labourers to fill 5 trucks daily (annual wages) | 96,000 |
| 3. Electricity & water charges | 18,000 |
| 4. Cost of Vermi castings for 20 beds @Rs.5/Kg (for 2500 kgs) | 12,500 |
| 5. Empty gunnies | 24,000 |
| 6. Watering charges (Per year) | 1,10,000 |
| 7. Cost of 100 tonnes of FYM @ Rs. 500/ ton | 50,000 |
| 8. Other miscellaneous charges | 18,000 |
| **Total recurring Expenses** | **3,88,500** |
| 9. Interest and Depreciation charges on non-recurring items @ 25% | 1,23,750 |
| **Total Costs** | **5,12,250** |
| **Revenue from 300 tonnes of Vermi compost @Rs 2/Kg** | **6, 00,000** |
| **Benefit-Cost Ratio = 6,00,000 : 5,12,250 = 1.17 : 1.00** | |

Source : Raw Data drawn from Venkataratnam.L and G. Purushottam reported in Mediplorama, 2000

A commercial scale vermi composting unit can give a return of 17 per cent on the investment.

## (B) Biofertilizers

Biofertilizers are microbial inoculants and they enhance crop production through improvement in nutrient supplies and their availability to crops. The potential benefits from using *Rhizobium* inoculation for some legume crops were estimated by Wani and Rao (1996) with the assumption that inoculation would increase yield by 10 per cent to 50 per cent of the area under the crop (Table 12.2).

**Table 12.2 : Potential benefits from using *Rhizobium* inoculation for selected legumes in India**

| Crop | Area (million ha) | Avg. Yield (Kg/ha) | Increased Yield (million tonnes) | Price per tonne US $ | Economic Benefits (million US $) |
|---|---|---|---|---|---|
| Chick Pea | 6.46 | 673 | 0.22 | 206.0 | 45.32 |
| Pigeon Pea | 3.63 | 663 | 0.12 | 235.0 | 28.24 |
| Groundnut | 8.35 | 1060 | 0.44 | 264.7 | 116.47 |
| Moong bean | 3.54 | 482 | 0.06 | 235.3 | 14.12 |
| Black gram | 3.61 | 521 | 0.09 | 233.5 | 21.18 |
| Lentil | 1.17 | 662 | 0.04 | 206.0 | 8.24 |
| Soybean | 3.63 | 856 | 0.15 | 200.0 | 30.00 |
| **Total** | | | **1.12** | | **263.57** |

Note: 1 US $ = Rs. 48.00
Source : Wani, S.P. and J.V.D.K. Kumar Rao, 1996

Wani and Rao also estimated the potential benefits from using $N_2$-fixing bacterial fertilizers for some selected cereal crops in India (Table 12.3). These estimates were based on the assumption that inoculation would increase yield by 10 per cent for sorghum and pearl millet, 15 per cent for rice, wheat and maize on 50 per cent of the area shown.

**Table 12.3 : Potential Benefits from $N_2$ -fixing bacterial fertilizers for selected cereal crops in India**

| Crop | Area (million ha) | Avg. Yield (Kg/ha) | Increased Yield (million tonnes) | Price per tonne US $ | Economic Benefits (million US $) |
|---|---|---|---|---|---|
| Sorghum | 12.9 | 900 | 0.58 | 88.2 | 51.2 |
| Pearl Millet | 9.5 | 530 | 0.25 | 88.2 | 22.0 |
| Rice | 42.0 | 1880 | 5.67 | 105.9 | 600.5 |
| Wheat | 24.9 | 2370 | 4.43 | 111.8 | 495.3 |
| Maize | 5.9 | 670 | 0.30 | 91.2 | 27.4 |
| **Total** | | | **11.23** | | **1196.4** |

Note: 1 US $ = Rs. 48.00
Source : Wani, S.P. and J.V.D.K. Kumar Rao, 1996

Although these potential benefits of biofertilizers in legumes and cereals appear to be very high, they will be realized only when the use of biofertilizers in these crops picks up and reaches 50 per cent of the area under these crops and the expected yield response of 10 to 15 per cent materializes.

Another important biofertilizer is blue green algae. It is recommended for use in rice crop. An estimate of its economic viability is presented in Table 12.4.

**Table 12.4 : Economics of Blue Green Algae Project (For a unit with production capacity of 50000 packets of 100 gms each)**

| | | |
|---|---|---|
| **A.** | **Project Cost** | **Rs. Lakhs** |
| 1. | Factory shed/building | 54.50 |
| 2. | Plant & Machinery | 10.00 |
| 3. | Fees for Technical know-how | 10.00 |
| 4. | Miscellaneous fixed assets | 1.67 |
| 5. | Preliminary and pre-operative expenses | 13.33 |
| 6. | Provision for contingencies | 7.95 |
| 7. | Margin money for working capital | 6.66 |
| | **Total** | **104.11** |
| **B.** | **Viability Indicators** | |
| 1. | Internal Rate of Return | 29% |
| 2. | Cash Break Even Point | 41.94% ( in first year ) |
| 3. | Debt Service Coverage Ratio | 1: 1.29 |
| 4. | Pay Back Period | 7 years |

Source : Technology Profiles circulated at Entrepreneurs' Meet on Biotechnology, Hyderabad, 1999

## (C) Bio-Agents

Bio-agents like *Trichogramma, Chrysopa, Coccinellids, Baculo viruses* etc., are being multiplied and supplied to the farmers for the control of pests. This is a highly capital-intensive activity requiring huge investments. The project economics of a bio-agents production facility are presented in Table 12.5.

**Table 12.5 : Project Economics of a Bio-agents Production Facility**

| | Project Cost | Rs. in Lakhs |
|---|---|---|
| 1. | Land and Site Development | 5.00 |
| 2. | Building | 24.00 |
| 3. | Plant & Machinery | 8.50 |
| 4. | Miscellaneous fixed assets | 8.00 |
| 5. | Technical know-how fees | 16.00 |
| 6. | Pre-operative expenses | 6.00 |
| 7. | Margin money for working capital | 1.50 |
| 8. | Contingency/escalation costs | 4.00 |
| | **Total** | **73.00** |
| 9. | Operating Cost | 125.00 |
| **10.** | **Turn Over** | |
| | (i) *Heliothis NPV* | 84.00 |
| | (ii) *Spodoptera NPV* | 84.00 |
| | (iii) *Trichogramma NPV* | 84.00 |
| | (iv) *Trichoderma* | 2.00 |
| | **Total** | **254.00** |
| **11.** | **Performance Indicators** | |
| | (i) Break Even Capacity | 40 tonnes |
| | (ii) Payback Period | 3 years |
| | (iii) Internal Rate of Return | 35 % |

Source : Technology Profiles circulated at Entrepreneurs' Meet on Bio-Technology, Hyderabad, 1999

## (D) Biopesticides and Plant origin Insecticides

Several biopesticides are in use today. Biopesticide products are based on natural agents such as microorganisms and fatty compounds. They are toxic to targeted pests and do not harm humans, animals, fish, birds and beneficial insects. One of the most common microorganisms used in biologically based pesticides is the *Bacillus thuringiensis* or Bt, bacterium. Several of the proteins produced by the Bt bacteria are lethal to individual species of insects. The European corn borer, one of the most prevalent pests, costs the United States $ 1.2 billion in crop damage each year. The biopesticides which are quite effective against this pest can cause a considerable saving even if they succeed in preventing 25 per cent of the losses.

As the interest on organic farming and integrated pest management is growing, plant origin insecticides like neem and tobacco formulations are gaining the acceptance of the farmers. During the last few years, there has been a marked increase in the use of neem formulations containing *Azadirachtin* as active ingredient both in India and abroad. In view of the growing demand, researchers are selecting superior trees (upto 0.8% *Azadirachtin* concentration) and propagating them through vegetative or tissue culture. A tissue culture protocol for large scale micro-propagation of neem was developed and standardised with the help of which a multiplication ratio of 1:10 in two and half months time was achieved. Farmers are planting these tissue culture neem clones in degraded lands with a hope to reap profits.

The economics of Neemgold 1500 ppm, a product of SPIC Ltd., Chennai are presented below in Table 12.6.

**Table 12.6 : Economics of Neemgold (1500 ppm *per litre)***

| | | Rs. in lakhs |
|---|---|---|
| 1. | Manufacturing cost | 101.55 |
| 2. | Market expenses | 24.00 |
| 3. | Sales Commission | 20.00 |
| 4. | Royalty | 2.00 |
| 5. | Excise duty and sales tax | 6.15 |
| | **Total Cost** | **153.70** |
| | Expected price | 200.00 |
| | **Net profit** | **46.30** |

Source: Technology Profiles circulated at Entrepreneurs' Meet on Biotechnology, Hyd, 1999.

## (E) Micro-propagation of Plants

Micro-propagation technology offers genetically identical plants popularly called as clones. The plants produced through tissue culture are free of diseases, pests and nematodes. Many companies are successfully producing tissue culture plants. The economics of a tissue culture project which can produce 3 million tissue culture plants per annum are presented below in table 12.7. This project requires a water supply of 25,000 litres per day and a power of 125 HP.

**Table 12.7 : Project Economics of a Tissue Culture Laboratory**

| A. | Project Cost | Rs. in lakhs |
|---|---|---|
| 1. | Land & Site Development | 8.00 |
| 2. | Factory Shed/Building | 45.00 |
| 3. | Plant & Machinery | 75.00 |
| 4. | Other Fixed Assets | |
| | (i) Green House | 15.00 |
| | (ii) Electrical | 15.00 |
| | (iii) Furniture & Fixtures | 3.00 |
| 5. | Know-how fee | 25.00 |
| 6. | Pre-operative Expenses | 20.00 |
| 7. | Margin Money for Working Capital | 8.00 |
| 8. | Contingencies and Location | 10.00 |
| | **Total** | **224.00** |

**Table 12.7 *Contd...***

| B. | Operative Cost | Rs. in lakhs |
|---|---|---|
| 1. | Consumables | 10.00 |
| 2. | Salary & Wages | 16.00 |
| 3. | Administrative Costs | 9.00 |
| 4. | Selling Expenses | 4.00 |
| 5. | Market Expenses | 6.00 |
| 6. | Power & Fuel | 15.00 |
| | **Total** | **60.00** |
| **C.** | **Economics** | |
| 1. | Cost of Production/plant at 80% capacity | 3.75 |
| 2. | Average Selling Price/Plant | 5.00 |
| 3. | Profit Margin/Plant | 1.25 |
| 4. | Internal Rate of Return | 25% |
| 5. | Payback Period | 5 years |

Source: Technology Profiles circulated at Entrepreneurs' Meet on Biotechnology, Hyderabad, 1999

## (F) Genetic Engineering

Genetically modified plants are created by the process of genetic engineering that allows scientists to move genetic material between organisms with the aim of changing their characteristics. In the absence of solid data on the impacts of modern biotechnology, some ex-ante evaluations are presented below.

Evenson (1994) evaluated the efforts to introduce insect resistance and disease resistance into rice under the Rice Biotechnology Programme of the Rockfeller Foundation, which was initiated in 1985. His estimates of the effects or benefits of Rice Biotechnology Programmes are presented in Table 12.8.

**Table 12.8 : Estimates of Effects or Benefits of the Rice Biotechnology Programmes**

| Trait or Benefit | Time to Production | | Area (million ha.) | Yield Increase (%) | Quantity (m. tonnes) | Value (Billion Rs.) |
|---|---|---|---|---|---|---|
| | Optimistic | Conservative | | | | |
| Multiple Insect Resistance | 12 | 21 | 37 | 30 | 41 | 8.0 |
| Multiple Disease Resistance | 15 | 22 | 50 | 15 | 27 | 5.4 |
| Total Annual Benefit | | | | | | 13.4 |

Source : Evenson, R.E., 1994

As can be seen, the estimated annual benefit of using insect-resistant and disease-resistant transgenic rice plants in farmers' fields is 13.4 billion dollars. The benefits are expected to start around the year 2012 (the mid-point between optimistic and conservative estimates). The Rockfeller Foundation's investment in the rice biotechnology programme between 1985 to 2012 would be approximately 0.3 billion dollars. The total financial support for rice biotechnology programmes from all other sources upto 2012 have been estimated to be about 2.4 billion dollars. Thus projected annual benefit, starting from the year 2012, would be approximately five times (13.4 / 2.7) larger than the total estimated support by Rockfeller Foundation and other sources between 1985 and 2012. A different way of calculating the annual benefit / cost ratio after the year 2012 is to divide $13.4 billion (total annual benefit) by $0.15 billion (estimated annual investment after the year 2012), which gives 90 as the benefit / cost ratio.

## (G) Pricing of Biotech Seeds

Let us assume that the 'Bollguard' cotton that Mahyco (Maharashtra Hybrid Seed Company) – Monsanto Research Foundation is presently testing in the farmers' fields passes through all stages of testing successfully and gets permission for commercial marketing. We further assume that it gives a 10% higher yield than the best ruling hybrid in all the irrigated cotton tracts, besides saving the cost of plant protection by Rs.1000 per hectare. The comparative economics are worked out and presented in Table 12.9.

**Table 12.9 : A Hypothetical Example of Benefits and Pricing of 'Bollguard' Cotton**

| Pricing | Ruling Hybrid (Rs) | Bollguard |
|---|---|---|
| Cost of seed (Rs/ha) | 1500 | - |
| Total cost of cultivation other than cost of seed (Rs/ha) | 15000 | 14000 |
| Total returns (Rs/ha)27000 | 29700 | |
| Farmers' profit before paying for seed (Rs/ha) | 12000 | 15700 |
| Farmers' profit after paying for seed (Rs/ha) | 10500 | - |
| Value addition due to Bollguard | - | 3700 |

Source: Rao K.P.C. , 2000

Given these hypothetical figures, the question arises as to at what level will Bollguard seeds, sufficient for planting one hectare, be priced? We can visualise several alternative scenarios.

1. The company may price them at Rs. 5200, thereby expropriating all value-addition due to their innovation.
2. The company may price them at Rs. 2500, if it figures out that the additional cost of research and development can be recovered by charging an additional price of Rs. 1000 per hectare over the present ruling price.

3. The company may price them at Rs. 3350, so that the value-addition due to new seeds is evenly shared between the company and the farmers who use the seeds.

Many skeptics think that the company may opt for scenario (i). In their perception, the company will fully exploit its monopoly to corner all benefits of innovation. But in reality, the company may not opt for this scenario because farmers will have no interest in using the new seeds if they do not get any benefit out of their usage. This will seriously constrain the demand for new seeds. Those who have experience with public sector agencies would wish that the company would opt for scenario (ii). But any company, which has taken risks and has invested money on an innovation, would not price its seeds merely on cost-recovery basis. The company may, instead, follow a middle path and opt for scenario (iii). It would enthuse the farmers to try the new seeds, as they will share about one-half of the benefit value-added by them. At the same time, the company will reap good returns on its investment, expand its sales and would have interest in further investing in the development of new bio-technology products.

But this scenario will last only for a short while. In the long run, as the new seeds are adopted by more and more farmers, the supply of cotton in the market would increase markedly, causing a fall in the price of cotton (may be not in absolute terms, but in real terms). That is the point when consumers will be getting their share of the benefit. But if cotton price falls, the value-addition due to innovation decreases and the company have to reduce its price. Thus, the new technology benefits all: the company, the seed growers, the farmers, and the consumers who use cotton-based products.

While the companies may come out with several new biotechnological products, all may not find acceptance with the farmers. For instance, the Roundup Ready (herbicide resistance) seeds of different crops may have very limited acceptance in India, because the use of herbicides itself is quite restricted in the country.

As more and more biotechnologies are released, their economic viability will be assessed based on micro-level data. *Ex-Poste* studies will grow in number, although *ex-ante* studies also have their relevance in case of emerging technologies. In any case, farmers will only adopt those biotechnologies, which are expected or believed to give a reasonable rate of return on the investments.

## II. CONCERNS ABOUT BIOTECHNOLOGY

The environmentalists, who are critical of biotechnology, have been raising several social, religious, ethical and biosafety issues, some of which are summarized below:

### 1. Ethical

(i) Should we alter the genetic structure of the entire living kingdom in the name of utility and profit?

(ii) Is there something sacred about life or should life forms, including humans, be viewed simply as commodities in the new bio-technological market place?

(iii) Do biotechnologists feel that they are masters of nature? Is this an illusion constructed on scientific arrogance and conventional economics ignoring the complexity of ecological processes?

(iv) Will some countries be plundered for their genetic resources?

### 2. Biosafety

(v) Is it possible to minimize ethical concerns and reduce environmental risks while keeping the benefits?

(vi) What will consumers be told about the new food products obtained through biotechnology (Right of the consumer to know)?

(vii) Will food from modified crops be safe?

(viii) Will food from modified plants have a different nutritional quality from that of the food it replaces?

(ix) Will modified plants transfer their introduced genes into wild relatives growing nearby?

(x) Could the planting of a restricted number of cultivars lead to a reduction in biodiversity and an increase in susceptibility to diseases?

### 3. Social

(xi) How will the structure of farming (particularly in developing countries) be affected by biotechnology?

(xii) How will patent laws affect the rights of traditional breeders? (For example, the right to save seed from one year to the next)

(xiii) What share of public resources (both financial and human) will be diverted to biotechnology research?

### 4. Economic Competition

(xiv) Is the genetic make up of all living things a common heritage of all or it can be appropriated by some Corporations and thus become a private property of a few?

(xv) Who gave individual companies the right to monopolise over entire group of organisms?

In view of the above concerns expressed by the skeptics of biotechnology, the role of the Government as the Regulatory Body attains importance. Some of the regulatory issues to be addressed by the Government are:

***Regulatory Issues***

1. Do current regulations give sufficient protection to farmers, consumers, those who have invested in research and those engaged in research?
2. Is there sufficient international legislation to ensure environmental protection?

3. Do current regulations compromise the competitiveness of biotechnology companies by being excessively restrictive?
4. Labelling of genetically modified crops
5. Financial investment by the government and industries in biotechnology.
6. Need to ensure transfer of technology to farmers and access to seeds.

## III. EQUITY ISSUES

Another debate that is taking place in the country is about the class of farmers that would benefit from the new biotechnology innovations. Some argue that only the rich farmers would benefit from the innovation because poor farmers will not have the ability to invest, nor the ability to take risk with a new technology. Some others argue that since the biotechnology innovations can be cost-saving, they would benefit the small farmers also. While these conflicting views are being voiced, the first group of analysts is apprehensive that the income distribution will worsen, and the second group of analysts expects that it would improve.

The experience of Green Revolution was that the rich farmers could reap the early gains from the new technologies. But subsequently, institutional mechanisms like custom hiring of tractors, water purchase from tube wells etc., made the new technologies accessible even to the resource-poor farmers and they also shared the benefits of technology, albeit, a little later. A similar trend may appear in case of biotechnology products as well. It is also established that whenever technology benefits the farmers, the demand for labour as well as the wage rates paid to them also increase. The secondary effects of technology contributed to the improvement of vertical income distribution. But, the Green Revolution technologies were noted to have increased the horizontal income inequalities between irrigated and rainfed areas. Will the same experience be repeated with biotechnological innovations?

Has biotechnology something to offer to people living in the ill-endowed regions? So far, biotechnology innovations have concentrated largely on insect-resistance and herbicide-tolerance. Ill-endowed regions like dry lands, saline-alkaline soils, degraded lands etc., would need tolerance to abiotic stresses to be built into seeds. The famous Dr. Anand Chakravarthy case relates to the invention of a microorganism that could eat up oil spills. Although there are no indications to that effect as yet, we may, by the same token, imagine microorganisms that can convert saline-alkaline soils into fertile fields. It may be mentioned that a Calcutta-based scientist recently claimed to have introduced successfully a salt-tolerant gene from the wild rice grown on the seacoast into cultivated varieties of rice. This success may be the first step in the evolution of high yielding rice varieties with the salt tolerance built into them. If such innovations materialize, biotechnology can reduce even horizontal income inequalities by aiding the development of backward regions and areas.

## IV. RELIGIOUS BELIEFS AND DIETARY RESTRICTIONS

As modern biotechnology involves transfer of genes across species and even from plants to animals and the *vice versa*, certain religious sentiments and dietary restrictions of the people may come in the way of adoption of some biotechnologies, even when they can yield high pay-offs to investments. Transfer of any animal genes to plants may cause an alarm to the strict vegetarians. Transfer of any genes from cows or pigs to plants or other animals may cause resentment among Hindus and Muslims respectively while consuming the modified food substances. But, vaccines or insulin developed from animal tissues are being accepted for medicinal purposes by the vegetarians. Going by this example, there may not be any opposition to GM foods on religious grounds. However, the mischief that fundamentalist forces of any hue or shade can play through orchestrated campaigns should not be underestimated. Time and experience can only tell whether or not the religious sentiments and dietary restrictions can hinder the acceptance of GM foods. The only way to blunt such campaigns is by way of education and informed discussions that an introduction of a single gene from somewhere would not alter the taste or composition of the food substances.

## V BUILDING PUBLIC AWARENESS AND DEBATE

In a democratic society, awareness campaigns and debates are necessary to appreciate the intentions and purpose of biotechnologists. Modern biotechnology has opened up the possibilities to reduce pollution and to improve the health and nutrition of human-beings. Even now, the functional literates constitute a small minority in the country. For centuries, people have lived with religious beliefs and superstitions. To make them appreciate the scientific inventions and discoveries, skillful campaigns are necessary.

### *Experiences with Green Revolution*

Over the last 54 years after independence, some headway was made in this regard. When Mashuri variety of rice was introduced, there was an apprehension that its consumption causes knee-joint pain. But after a decade, consumers have given it the top preference for consumption. How the initial objection was overcome ? Just by education and experience. Consumers also believed that food grown with chemical fertilizers does not have the same nutritive value as that grown with organic manures. But, later this resistance was overcome. Of course, now the issue has completed a full circle with the quality experts and rich consumers preferring organically grown food that does not have any residues of pesticides. The farmers and consumers may also accept GM foods in the same way as long as it is proved that they do not have any deleterious effects on health and nutrition of humans, animals and plants. The reduction in pesticide use should lead to a lower level of residues and to a positive preference of GM foods, if lasting resistance to pests and disease can be built into plant varieties through modern biotechnology.

Social scientists, who should contribute to the debate and public awareness of GM foods, have to adopt as neutral a stance as possible. Like any other section of the society, some of them may tend to sweep aside the risks in order to accelerate the materialization of benefits to the public, while some others may exaggerate the risks and downplay the benefits of modern biotechnology. Despite such extreme positions taken by some individual social scientists, the overwhelming majority of social scientists are expected to take a dispassionate and objective view of this contentious issue.

### *Better Chances of Acceptance of GM foods*

In less developed countries like India, people still spend a considerable part of their income on food (about 30 to 50 per cent). In the developed countries of Europe, food expenditures are below 10 per cent of the total expenditures. If the genetic engineering technologies can reduce the unit cost of food substances by 10 per cent, it may not mean much to European consumers, but the same can improve the access to food in case of consumers of developing countries, particularly in case of those below the poverty line. Poor people are known to take greater risks for survival and development than the rich. In view of these well-known facts, the GM foods may be accepted better in developing countries, if it is proved that they have both cost-reducing and health-enhancing properties. Notwithstanding these stylized facts, a proper assessment of safety of GM foods is absolutely necessary. Both the testing processes and science-based debates should be as transparent and impartial as possible.

### *Infrastructure for Biosafety Tests*

Farmers' Organizations, Consumer Associations and Non-governmental Organizations should also be involved with the implementation of effective biosafety protocols. Presently, we are relying on toxicology institutes for this purpose. We need separate institutions to look at toxicology and biosafety aspects of agricultural products. We have developed such protocols for drugs and to some minor extent for industrial pesticides and chemicals but nothing at all for biological products. Today, the drug industry in developed countries produce 60000 to 80000 new molecules every month that go through 30 different tests. They are screened cell-based and each of them costs about 20 to 30 million dollars to buy. If the Indian Council of Agricultural Research (ICAR) or State Agricultural Universities buy those Cell-screening Systems, large numbers can be screened very quickly, instead of using animal systems and then wait for an extended period of time to get the final results. India also lacks a smooth and fast protocol for field trials of transgenic plants before they can be released commercially. Similarly, we don't have adequate standards for animal biotechnology.

After strengthening infrastructure and protocols for a speedy and comprehensive assessment of biosafety, the results are to be shared widely with the stakeholders. Of course, there is one problem in case of biotechnological innovations. With the strengthening of intellectual property rights (IPRs) as required by the World Trade Agreement (WTA), there is a lot of awareness and interest in patenting the products and processes rather than in publishing the results. Of course, disclosure is a pre-condition for grant of any forms of IPRs. But the lure of commercial exploitation of new products and processes may delay the availability of results to the common public. Not withstanding this limitation, there is a continuous need to publicize the results as early as possible to clear the misconceptions and prejudices of the public built on rumour and ignorance.

### *Dissenting voices of Environmentalists*

Some self-styled environmentalists have seen violence in the green, blue and white revolutions ushered in the past few decades. But it is these revolutions that have averted a 'red' revolution by increasing the physical food availability faster than the population growth and by making it economically accessible to the vulnerable sections of population through the reduction in the real costs of food articles. It is an irony that scientists responsible for green, blue and white revolutions are remembered and admired much less than these self-styled champions of peoples' interests. Two centuries of slavery and the resultant scourges of poverty, illiteracy, inequality and dependency have provided fertile grounds for sowing the seeds of doubt by these professional agitators to reap the benefits of fame and recognition. This is not to ignore or to sweep aside the problems of resource and environmental degradation that have assumed alarming proportions because of wrong priorities and deferred investments. In fact, considerable resources should be allocated to rejuvenate the basic resources and environment to their past glory. But in this effort, we should use all the advances in science and technology to keep up the increased and diversified supplies of food and other articles while simultaneously restoring resources to the pink of their health. In a resource – scarce country like India, we can ill afford to ignore the cutting edge technologies that are advancing by leaps and bounds in other parts of the world.

### *Fear of the Multi-National Corporations*

For a Nation fed on the slogans of socialism and public sector, private sector and profit are not decent words. But we have long realized that the public sector failed us and that the egalitarian society was only a dream that was never achieved. The new economic policies of liberalization, privatization and globalization are getting a fair trial. In spite of some problems caused by the World Trade Agreement, the country is learning to face, negotiate and grapple with the issues related to international trade. The entry of multi-national corporations either by themselves or in partnership with local firms is a fact that we should reckon with. History is replete with examples that the social returns from technologies are much higher than the private returns expropriated by any form of private organization. Of course, monopolies resulting from mergers and acquisitions are a cause of concern. We should try to increase competition in the game of invention by investing heavily in the public sector laboratories and in training and encouraging the promising scientific workers in the field of Biotechnology. Strengthening of Intellectual Property Rights would motivate and aid the processes of technology generation. While emphasizing on the need for adequate testing of biotechnology outputs, we should not develop a prejudiced view against the private companies or, for that matter, multinational companies. The government should ensure that the free choices of producers and consumers would prevail. Safe products of biotechnology research, if found to be having the potential to increase output and to reduce cost and pollution should be welcomed irrespective of the source of their origin.

## VI. KEY MESSAGES

Some key messages that should find a place in the public awareness campaigns for scientists, policy makers, NGOs, extension workers, farmers and consumers may be summarized as follows:

1. Science and technology have unlimited potential to solve the problems plaguing our society.
2. Human societies have always opposed new discoveries due to the fear of unknown.
3. Nothing should be pre-judged based on prejudice or fear.
4. Technologies that perform well will eventually be accepted by the society.
5. Adequate and comprehensive testing of new products is necessary.
6. Stakeholders have a right to know all the pros and cons of a new technology.
7. Healthy debate leads to a resolution of conflicting perceptions and to the emergence of a consensus.
8. Social and moral concerns may limit the acceptance of new technologies.
9. Objective facts will eventually score over subjective beliefs.

10. Social returns from technologies are much higher than private benefits.
11. Priority setting in research can lead to a better focusing on the needs of the poor.
12. All sections of society will eventually benefit from new technologies, although they may be received by a particular section in the first instance.
13. A liberal and democratic society is the best bet for informed discussion and rational choices.
14. Regulatory bodies should carry out their jobs in a strict, impartial and transparent manner.
15. *Ex-Ante* and *Ex-Poste* evaluations of new technologies are needed for rational decisions.
16. Competition and rewards are necessary to support inventive activity.
17. Competitiveness has to be improved along with quality of products.
18. By developing legal and institutional safeguards, conflicts in the interests of different sections can be resolved.
19. Socio-cultural development of have-nots is as much important as their economic empowerment.
20. Widespread validation is needed for adoption and adaptation of technologies by the farmers.

## VI. STRATEGIES TO REACH TARGET GROUPS

In a differentiated and multi-layered society like India, a combination of strategies is required to take the messages to all sections of the population. Some of them are :

1. Wide spread testing and validation of research results is required to evolve technologies relevant for different sections of the farmers.

2. Demonstrations of proven technologies are necessary to convince the farmers.
3. Awareness campaigns about new but proven methods through mass media are necessary to develop interest in them.
4. Subsidizing organic inputs and promotion of environment-friendly methods of production will help in early adoption of new technologies.
5. Training and skill development can have high pay-offs in the long run.
6. Producers / consumers associations and clubs to discuss and debate about new technologies.
7. Farmers can be involved in research initiatives in a partnership or participatory mode.
8. Contractual arrangements between companies and farmers to ensure supply of quality inputs and processing facilities.
9. Non-governmental organizations to complement the efforts of governmental agencies in public awareness campaigns and debates.

### *References*

Biotechnology : Reaching the Unreached – Dialogue Proceedings No : 3 of M.S. Swaminathan Research Foundation, Madras, January, 1991.

Biotechnological Interventions for abiotic Stress Management in Rainfed Crops – Pre-Project Formulation Workshop Organized by Biotechnology Unit, IPE, Hyderabad at CRIDA, Hyderabad-59 from July 11-13, 2000.

Evenson, R.E. – 'An *Ex-Ante* Economic Evaluation for the Rice Biotechnology Programme' in "Setting Priorities for Rice Research" (Hossain M, R. Herdt and R.E. Evenson, eds), 1994.

Federation of Indian Chamber of Commerce and Industry – Proceedings of the Seminar on 'Agro-Processing and Biotechnology' held at Bangalore on July 20, 2001, New Delhi.

Genetically Modified Plants : Implications for Environment, Food Security and Human Nutrition – Recommendations of the Participants of National Consultation, M.S. Swaminathan Research Foundation, Madras, January 6-7, 1999.

Kendall, Henry, W, *et al* – Bio-engineering of Crops – Report of the World Bank Panel on Transgenic Crops, The World bank, Washington D.C: 1997.

Mediplorama 2000 – Convention on Medicinal and Aromatic Plants, May 19-20, 2000, Hyderabad – 38.

Paarlberg, Robert L – 'Governing the GM crop Revolution', International Food Policy Research Institute, Washington D.C: 2000.

Prakash, C.S. – 'The Relevance of Biotechnology to Agriculture in India' Rural Banker, Issue No: 24, February 2000, pp 11-18

Rao, K.P.C. – 'Economic Implications of Commercializing Transgenic Plants' in Proceedings of the Workshop on "Stakeholders' Dialogue on Agricultural Biotechnology: Biosafety and Economic Implications" Organized by Tata Energy Research Institute at TNAU, Coimbatore on November 25, 1999, New Delhi: 2000.

Rao, S. R. – 'Building Public Acceptance of Genetically Modified Products in India' Presented at the International Conference on "Biotechnology for Sustained Productivity in India" Organized by International Life Sciences Institute, India on November 1-2, 1999 at Hyderabad.

Regional Biofertilizer Development Centre – 'Biofertilizers Scenario of Andhra Pradesh' – A Documentary Report, Nagpur: 2000.

Sharma, N and P.S. Janaki Krishna – 'Status of Plant Biotechnology in India', Institute of Public Enterprise, Hyderabad: 1999.

Technology Profiles – Entrepreneurs' Meet on Biotechnology, Organized by APIDC, APSFC, Industries and Commerce Department, Government of Andhra Pradesh and Biotech Consortium India Limited, November 26, 1999, Hyderabad.

Wani, S.P. and J.V.D.K. Kumar Rao – 'Biofertilizers: Potential and Constraints in Dry land Agriculture' in Proceedings of the Pre-project Formulation Workshop on "Biofertilizers and Biopesticides" Organized by Biotechnology Unit, IPE at Krishi Vigyan Kendra, Gaddipalli on September 30 and October 1, 1996.

# 13

# Indian Rules and Developments in Transgenic Organisms and Cartagena Protocol on Biosafety

* P. K. Ghosh

The Government of India enacted[Ref 1] its Environment (Protection) Act (EPA) in 1986. The Act was to be the instrument for preserving Indian environment and to minimize the risks from pollutants and contaminants that could affect its flora and fauna, taking also into account the risks to human and animal health. It was the expectation that the Genetically Modified Organisms (GMOs) were to play important role in the economic uplifting of the country in its various facets of applications. These include human and animal health care area, agriculture, industrial products and environment management. Concurrently it was also realized that there could be unintended hazards and risks from the use of GMOs and products

* Ex Advisor, Department of Biotechnology, Ministry of Science & Technology, C2B/5A, Janakpuri, New Delhi – 110 058.

Views expressed in this paper are those of the author and they do not necessarily express the views of the organization to which the author belongs/belonged.

thereof, if the new technology was not properly assessed before use. In order to produce a GMO, a gene construct comprising a host-compatible-promoter, a gene of interest and a terminator sequence or a polyadenylation sequence is integrated in a stable manner into the genome of an organism for the target gene to be expressed and stably inherited. A GMO can be safe but this can be unsafe too. This will depend upon the trans-genes, the host organism and the environment where the GMO is being tested. Transgenic rape seed plants containing *bar* genes transferred the transgenic trait to near relatives of *Brassica* spp. when such seeds were mass multiplied in commercial agriculture in Canada. Insect resistant *Bt* plants coding for specific *Bt* proteins developed *Bt* protein resistant insects in laboratory experiments. Transgenic soybean genetically modified to increase its sulfur containing amino acids by incorporating Brazilian nut 2S gene was allergenic to serum of people who were allergenic to Brazilian nut 2S protein. Potatoes genetically modified with specific lectin genes protected attack from insects but such potatoes were not safe to rodents when they were fed with such potatoes. The transgenic pollens of corn coding for *Bt* proteins killed the monarch butterfly larvae when they were forcibly fed with such pollens. It is expected that transgenic pollens coding for *Bt* proteins would affect the silkworm larvae, as these are insects that are susceptible to *Bt* proteins. There are examples of microorganisms, especially genetically modified viruses that turned virulent after modification. The longevity of GM fish was shortened, compared to the non-transgenic controls. Consequently, a case-by-case analysis of the safety of each GMO needs to be conducted to assess environmental safety as well as safety to human and animals. Keeping the potential adverse effects in view, the Indian Government had issued Rules & Procedures (Rules) for handling GMOs and hazardous organisms through a Gazette Notification No. G.S.R. 1037(E) dated 5.12.1989 from the Union Ministry of Environment & Forests[Ref 2]. The Rules cover all kinds of GMOs and products thereof, which are controlled commodities for handling and use in the country under the EPA.

The Convention on Biological Diversity (CBD) which is a convention created by several countries from all over the world in 1992 in Rio, Brazil in its Article 19(3) established the mandate for the negotiation of a Protocol on Biosafety, emanating from the use of GMOs. The GMOs have been designated as Living Modified Organisms (LMOs) in the documents of CBD. The above Article required the Parties (Signatory Members States to the CBD) to consider the need for and modalities of a protocol setting out appropriate procedures, including, in particular, advance informed agreement, in the field of the safe transfer, handling, and use of any living modified organism resulting from biotechnology that may have adverse effect on the conservation and sustainable use of biological diversity. Article 19(4) of CBD creates an obligation for Parties to the CBD to provide information on any LMO transferred to another Party. The obligation exists in the Convention independently of the Protocol. Together, CBD Article 19 (3) and (4) constitute the provisions of the CBD in dealing with the establishment of international standards on LMOs. Article 8 (g) of the CBD requires Parties to regulate, manage or control domestically the risks associated with the use and release of LMOs resulting from biotechnology, if the conservation and sustainable use of biological diversity, taking into account the risks to human health, could be affected. It is interesting to note that this paragraph of the article refers to human health also. Article 17 of CBD deals with exchange of information; a reference to this article in the Preamble to the Protocol underlines the importance of information sharing in as far as bio-safety is concerned, as well as the special needs of developing countries in this field, which Article 17(1) specifically refers to. The Conference of Parties to the CBD, at its second meeting in November 1995, considered the need for a Protocol on Biosafety, as required by Article 19(3) of the Convention. Its decision to develop such a Protocol (Decision II/5) constitutes a milestone in the history of the development of the Biosafety Protocol. The Decision provides guidance to the CBD Parties in the development of the text. The text of the Biosafety Protocol was adopted in January 2000 by the Conference of Parties in their meeting in Montreal. The objective of the Protocol is to contribute to ensuring an adequate level of protection in the field of safe transfer, handling and use of living modified organisms resulting from modern biotechnology that may have adverse effects on the conversation and sustainable use of biological diversity, taking also into account risks to human health, and specifically focusing on trans-boundary movements.

## Indian EPA Implementation Structure for GMOs

The Indian EPA Rules define Competent Authorities and compositions of such Authorities for the handling of all aspects of GMOs and products thereof. The GMOs include microorganisms, plants and animals. Presently, there are six competent authorities as stated below, indicating their broad responsibilities and authorities too:

### (i) The Recombinant DNA Advisory Committee (RDAC)

This Committee constituted by the Department of Biotechnology of the Union Ministry of Science & Technology is to monitor on the developments in biotechnology at National and International levels. The RDAC submits recommendations from time to time that are suitable for implementation for upholding the safety regulations in research and applications of GMOs and products thereof. This Committee prepared the first Indian Recombinant DNA Biosafety Guidelines in 1990, which was adopted by the Government for conducting research and handling of GMOs in India[3].

### (ii) The Review Committee on Genetic Manipulation (RCGM)

The RCGM is constituted by the Department of Biotechnology to monitor the safety aspects of ongoing research projects and activities involving genetically engineered organisms. The Committee is also mandated to bring out Manuals of Guidelines specifying procedures for regulatory process with respect to activities involving genetically engineered organisms in research, use and applications including industry, with a view to ensure environmental safety. All ongoing projects involving high risk category and controlled field experiments shall be reviewed by the RCGM to ensure that adequate precaution and containment conditions are followed. The RCGM can lay down procedures restricting or prohibiting production, sale, importation and use of GMOs.

RCGM can approve applications for generating research information on transgenic microorganisms. The growth of microorganisms under sterile conditions in a fermentation vessel under submerged conditions can be carried out in large bioreactors. However, RCGM can approve experiments in bioreactors having geometric volume of up to 20 litres. All experiments for the use of larger bioreactors require the approval of the Genetic Engineering Approval Committee (GEAC). RCGM puts conditions of contained use of the bioreactors in order to prevent the escape of genetically modified microorganisms into the open environment. The procedures for safe handling of microorganisms[Ref 3] have been stated in the Recombinant DNA Safety Guidelines -1990·

RCGM can also approve applications for generating research information on transgenic plants. Such information can be authorized for generation in contained green house as well as in small plots. The small experimental field trials are limited to a total area of 20 acres in multi-locations in one crop season. In one location where the experiment is conducted with transgenic plants, the land used should not be more than one acre. The design of the trial experiments requires the approval of the RCGM. The design of the experimental plot in open environment is made to seek answers to relevant and necessary questions on environmental hazards including risks to animal and human health. Data are required to be generated on economic advantage of the transgenics over the existing non-transgenic varieties. RCGM can also direct the generation of toxicity, allergenicity and any other relevant data on transgenic materials in appropriate systems.

For generating research information on transgenic animals, RCGM can authorize the investigator to conduct experiments in the lab as well as in contained and enclosed conditions so as to prevent the escape of the transgenic animals into the open environment. The experiments are designed to generate information about the growth characteristics and the health conditions of the transgenic animals, using the non-transgenic animals as controls.

The RCGM can issue clearances for import/export of etiologic agents and vectors required for producing and cloning genetically modified microorganisms, plants and animals. Clearances for import/export are also provided for transgenic microorganisms; transgenic germplasms including transformed calli, seed and plant parts; as well as transgenic animals of various kinds for research use only. All experiments using GMOs, which belong to category III risks and above, as elaborated in the guidelines require a permit to be issued by the Department of Biotechnology for authorizing such experiments. All such permits are issued on the basis of the recommendations of the RCGM. According to the Indian classification of risks, the category-I risks involve routine recombinant DNA experiments in lab and work involving defined genes/DNA of microbial, plant or animal origin which are generally considered as safe. Category-II risks involve lab and contained greenhouse experiments involving genes or DNA of microbial, plant or animal origin, which are non-pathogenic to human, but can have implications on plants and insects. Fermentation experiments with GMOs conducted in fermentation vessels up to 20 litre geometric volume can be Category-II risk experiments, if the transgenic microorganisms or cell lines used are harmless and non-pathogenic. Such experiments can also belong to Category-III risks, if the GMOs are coding for toxins or are infective to human and animals. Other Category-III risk experiments involve genes/DNA of microbial, plant or animal origin, which can cause alterations in the biosphere and does not fall in category-I & II. All open field experiments of GMOs howsoever organized, are considered to belong to category-III risks although they may be carried out under reasonably contained conditions by taking all the precautions to prevent the escape of GMOs or parts thereof that have propagating traits into the uncontrolled open environment.

The RCGM can put such conditions as would be required to generate long-term environmental safety data from the applicants seeking release of transgenic microorganisms, plants and animals into the open environment. RCGM can also setup expert committees to monitor the research experiments. For monitoring the contained field experiments with GM plants, the RCGM had setup a Monitoring-cum-Evaluation Committee (MEC) with many agricultural experts as its members. MEC makes on the spot visits of the experimental sites and advises the RCGM about the steps to be followed in the conduct of experiments for assessing agronomic benefits, besides conducting environmental risk assessments.

The RCGM had revised the 1990 Guidelines for conducting research using GMOs in May, 1994 [Ref 4] and subsequently in August, 1998 (incorporating further amendments in September, 1999) [Ref 5]. The present guidelines have emphasis on genetically modified microorganisms and plants. The latest guidelines capture detailed procedures for conducting contained field experiments using GM plants; they also provide guidance for generating food safety data for transgenic plants. Contained field experiments are designed to account for and to arrest the escape of the transgenic plants or plant parts or seeds set in the plants into the open environment; they also designed to create reasonably effective barrier to prevent the escape of the transgenic pollen into the open environment. The author in detail[Ref 6] discussed Indian structure on biosafety of transgenic plants earlier.

**(iii) Institutional Biosafety Committee (IBSC)**

This Committee is constituted by organizations involved in research with GMOs. The Committee requires the approval of the Department of Biotechnology. IBSC also has a nominee from the Department of Biotechnology who oversees the activities to ensure that safety aspects in accordance with the safety guidelines are fully adhered to by the organization. Every R&D project using GMOs has to have an identified investigator who is required to inform the IBSC about the status and results of the experiments being conducted. Experiments belonging to Category I and II risks as well as all experiments conducted with GMOs in the contained lab, contained green house conditions for plants as well as contained lab, caged or enclosed conditions for animals can be approved by the IBSC; however, the synopsis of all such experiments is required to be reported to the RCGM in the Department of Biotechnology in the form of reports from time to time in a prescribed format. Such information along with the progress of research work is also reported to the RCGM as a mandate at least once in six months.

### (iv) Genetic Engineering Approval Committee (GEAC)

This Committee functions as a body in the Ministry of Environment & Forests and is responsible for approval of activities involving large-scale use of GMOs in research, industrial production and applications. The clearance of GEAC is only from environmental angle under the EPA. All other relevant laws would apply even though EPA clearance is available for using GMOs and products thereof; for example, drugs made through GMOs would require separate approval for manufacture and use under the Indian Drugs Act; production of GMOs is also authorized under Indian Industries (Development & Regulation) Act, and therefore these clearances are also mandatory.

Large-scale experiments beyond the limits specified within the authority of RCGM are authorized by GEAC only. The GEAC can authorize approvals and prohibitions of any GMOs for import, export, transport, manufacture, processing, use or sale under Rule 7,8,9,10 & 11. All such authorizations are usually conditional, and Rule 13 guides such conditions.

### (v) State Biotechnology Coordination Committee (SBCC)

This Committee, headed by the Chief Secretary of the State is constituted in each Indian state where research and applications of GMOs are contemplated. The Committee has the powers to inspect, investigate and take punitive actions in case of violations of the statutory provisions. The Committee coordinates the activities related to GMOs in the State with the Central Ministries. This Committee also nominates State Government representatives in the activities requiring field inspection of activities concerning GMOs.

### (vi) District Level Committee (DLC)

This Committee constituted at the district level is considered to be the smallest authoritative unit to monitor the safety regulations in installations engaged in the use of GMOs in research and applications. The District Collector heads the Committee who can induct representatives from State agencies to enable the smooth functioning and inspection of the installations with a view to ensure the implementation of safety guidelines while handling GMOs, under the Indian EPA.

## Some relevant paras of Rules 1989 on GMOs

There are 20 paras in the Rules 1989. Different paras deal with different aspect of the Rules. Each para can also be designated by suffixing the number of the para and by prefixing the word Rule for identifying them.

Para 7 (can also be called as Rule 7 as explained above) of the Rules deal with approval etc. by individuals on the import, export, transport, manufacture, process, use or sale of GMOs. This para deal with use of GMOs for the purpose of researches in laboratories that are notified by the Ministry of Environment & Forests. Further, it directs that the GEAC authorize individuals and laboratories to take proper measures for the handling or discharge of GMOs in the open environment. Use of GMOs in large plants as well as in pilot plants requires a license from the GEAC. Certain experiments of lower risk could be carried out with the approval of the IBSC. This para does not specifically state what steps are to be taken for experiments conducted in the small scale under fully contained conditions. Indeed, the powers of the GEAC as embodied in para 4(4) talk about authorization of GEAC for activities involving large-scale use of GMOs in research and industrial production. There is no clear-cut indication of how contained small-scale research of GMOs is to be dealt with. On interpretation it appears that this is within the purview of the RCGM as per Rule 4(2). Realizing this situation, internal working arrangements have been made to allow RCGM to handle contained small-scale research using GMOs.

Para 8 deals with production of GMOs where authorization for production is to be obtained from the GEAC. GM microorganisms, plants and animals require authorization for commercial use in accordance with this para.

Para 9 deals with deliberate or unintentional release of GMOs into the open environment. For this purpose, all situations of use are to be authorized by the GEAC.

Para 10 and 11 deal with approval for substances, which may contain GMOs. In such cases also where substances, such as, food products are thought to be containing GMOs, the use of such products in commercial arena requires authorization from GEAC.

Para 12 deals with procedures for applicants for obtaining approval for use of GMOs under different conditions. The applicants are required to make an application in a prescribed format.

Para 13 deals with conditions of approval of GMOs for commercial use. It can be seen from the recital of para 13 that Government applies a precautionary principle while granting permission for marketing any GMO or products thereof. All commercial authorization is for a limited period, which requires renewal after the expiry period. Further, approvals are also given with conditions of observing and collecting information from the country on the risks, if any, arising from the commercial use of GMOs and products thereof.

Para 14 deals with a mechanism for supervising the implementation of the terms and conditions under which approvals for the marketing and commercial use of GMOs and products thereof are authorized.

Para 15 deals with penalties that can be levied/imposed on persons/ institutions/companies who are responsible for non-compliance of measures, which requires to be taken for the safe use of GMOs and products thereof.

The interrelationship among the various authorities, and the applicable Rules can be better comprehended by referring to Fig. 13.1.

## Scope and Procedures of Cartagena Protocol on Biosafety and Indian laws :

There is a need to compare how the Cartagena Protocol on Biosafety compares with the Indian Environment (Protection) Act and Rules. The Table 13.1 below gives the comparison between the two legal structures.

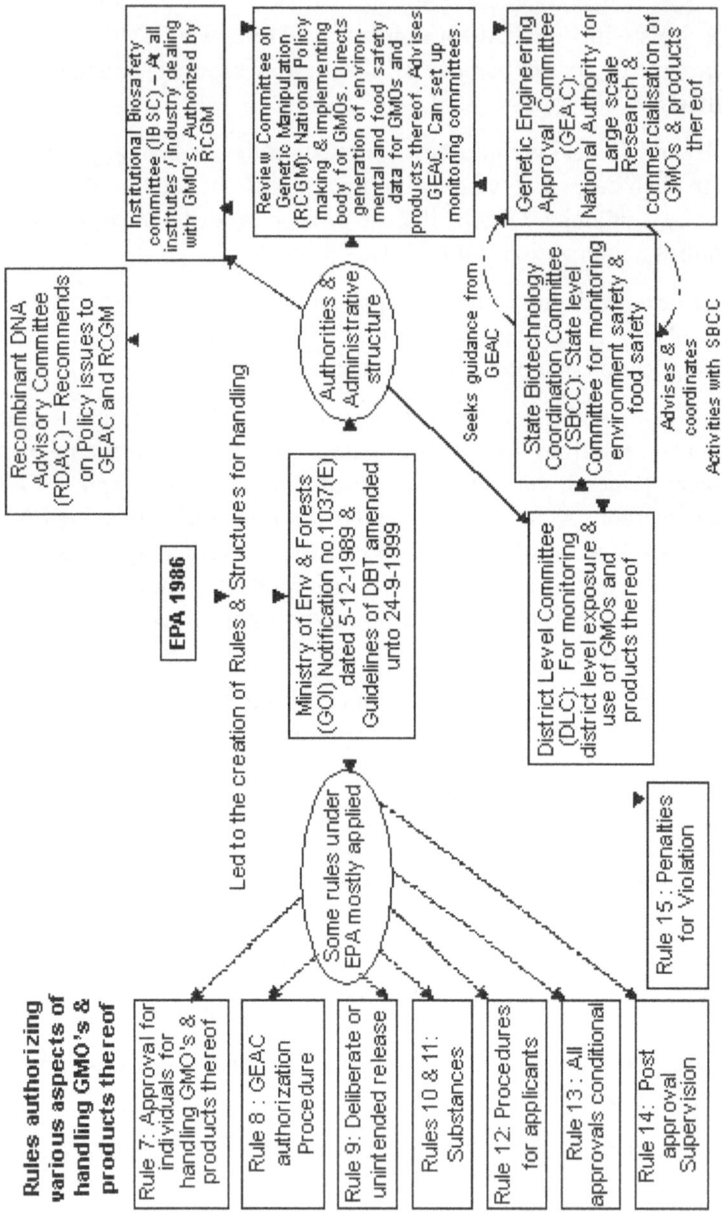


**Fig. 13.1** The Indian Rules, Authorities and Administrative Structures for Dealing with GMOs and Products thereof

**Table. 13.1 Scope and procedures of cartagena protocal on biosafety and Indian laws**

| | Cartagena Protocol on Biosafety | | Indian Environment Protection Act And Rules | | Remarks |
|---|---|---|---|---|---|
| | **Preamble / Article** | **Requirements / Tasks / Actions / Procedures** | **Rules / Para** | **Requirements / Tasks / Actions / Procedures** | |
| 1. | Preamble | Reaffirms precautionary approach contained in Principle 15 of the Rio declaration. | Preamble | Affirms protecting the environment, nature and health while using Genetically Modified Organisms (GMOs) or products thereof | Indian Act & Rules deal with Genetically Modified Organisms (GMOs) and products thereof on a precautionary principle while Cartagena Protocol deals with Living Modified Organisms (LMOs) & affirms precautionary approach. The distinction of treatment for & between the living and non-living organisms, both genetically modified, can be noted as difference between the Indian Rules and the Cartagena Protocol. |

Table 13.1 *Contd.....*

| | | | | | |
|---|---|---|---|---|---|
| 2. | Art. 2(2) | To ensure the development, handling, transport, use, transfer and release of LMOs are undertaken in a manner that prevents or reduces the risks to biological diversity, taking also into account risks to human health. | Rules 2(1), 2(2), 2(3), 2(4) & 2(5) | | Applicable to manufacture, import, storage, sale, offers for sale and any kind of handling of GMOs and products thereof. Though Indian Rules do not specifically affirm the use of precautionary principles for handling GMOs and products thereof, they affirm by implication the application of precautionary principle through the narrations as stated in the preamble. |
| 3. | Art. 8(2) & 11(2) | To ensure that there is a legal requirement for the accuracy of information provided by domestic exporters for the purpose of notifications for export to another country and by domestic applicants for domestic approvals for LMOs that may be exported as LMO-FFPs. | Rule 2(4) & 12 | All kinds of handling including exports and imports, production manufacture use and storage require submission of information. No distinction is made particularly for LMO – FFPs. | The Genetic Engineering Approval Committee (GEAC) under the Indian Rules has prescribed formats for submission of information from the applicants who apply for approvals for manufacture cell import or use of GMOs and product thereof. The information has to be accurate and to this effect the applicants have to give a declaration. |

**Table 13.1** ***Contd.....***

| | | | | | |
|---|---|---|---|---|---|
| 4. | Art. 8, 9(3) & 10(1) | The Party of Export (POE) shall notify or require the exporter to ensure notification to the & 10(1) Party of Import (POI) on any trans-boundary movement of LMOs and the notification shall at a minimum contain information in accordance with those prescribed in Annex. 1 of the Protocol. The POI can ensure that domestic regulatory framework used in place of the Advanced Informed Agreement (AIA) procedures are consistent with the Protocol. POI can further ensure that AIA decisions are taken in accordance with Art.15. | Rule 7 (1), (2) ,(3), (4) & (5) and Rules 8 – 11 | The Indian Rules require that applicants seeking to import, export, transport, manufacture, process, use or sale GMOs or products thereof furnish information of risks assessment in a prescribed format to the satisfaction of the GEAC. | Indian Rules do not distinguish GMOs for export purposes as different from GMOs used domestically. However all authorization of GMOs and products thereof for commercial use requires approval from the GEAC. The later ensures that authorization is accorded only after risk assessment is carried out in accordance with Rules. There is presently no procedure for notifying the importer in Indian Rules. |

**Table 13.1** *Contd.....*

| | | | | | |
|---|---|---|---|---|---|
| 5. | Art. 15(1), (2) | To ensure that risk assessments are carried out for decisions taken under Article 10 and that they are carried out in a scientifically sound manner. | Rules 4(2), 4(3) & 4(4) | Risks assessments for category I & II risk experiments are carried out under the supervision of the IBSC. For higher category risk RCGM followed by GEAC conducts risk assessment studies. | The GEAC before releasing any GMOs or products thereof into the open environment ensures that adequate risk assessment information has been generated. |
| 6. | Art. 16(1), 16(3) 16(4) | Within the provisions of the protocol to establish and maintain appropriate mechanisms, & measures & strategies to use, handling and transboundary movement of LMOs, to take measures to prevent unintentional transboundary movements of LMOs and to ensure that LMOs whether imported or locally developed, have undergone an appropriate period of observation that is commensurate with its life cycle or generation time before it is put to its intended use. | Rules 14(1) 14(2) 16(1) & 16(2) 17(1) & 17(2) | The GEAC would supervise the implementation of the terms and conditions laid down in connection with the approvals accorded by it for environmental release of GMOs or products thereof. | The GEAC through the RCGM can cause to conduct any number of experiments to ensure that safety related questions concerning the use of GMOs or products thereof are adequately addressed. |

Table 13.1 *Contd.....*

| | | | | | |
|---|---|---|---|---|---|
| 7. | Art. 17(1) | To take measures to notify affected or potentially affected states, the Biosafety Clearing House, and , where appropriate, relevant organizations, when there is an occurrence within its jurisdiction that leads or may lead to an unintentional trans-boundary movement of LMOs that is likely to have significant adverse effects. | ----- | ----- | Indian EPA Rules have jurisdictions within the Indian territory only. Therefore, the Rules do not address concerns in other countries. |
| 8. | Art 18(1), 18(2a), 18(2b) & 18(2c) | To take measures to ensure that LMOs that are subjected to tans-boundary movement under the protocol are handled , packaged and transported under conditions of safety, taking into consideration relevant international Rules and standards. | Rule 1(1), 2(1), 2(3), | The Rules define the coverage of LMOs and products thereof. | The Indian Rules do not specifically address matters related to handling, packaging & transport of LMOs and products thereof. However, the RCGM has specified conditions related to handling, packaging and transport of GMOs as well as products thereof for research materials in order to ensure that such materials do not easily reach the open environment. |

**Table 13.1** ***Contd.....***

| 9. | Art.<br>21(1)<br>21(2)<br>21(3)<br>21(4)<br>21(5)<br>21(6) | To provide for the designation of confidential information by notifiers, subject to the exclusions set forth in Art. 21(6). To ensure consultation with notifiers and review of decisions in the event of disagreement regarding claims of confidentially. To ensure the protection of agreed-upon confidential information and information claimed as confidential where a notification is withdrawn. To ensure that confidential information is not used for commercial purposes without the written consent of the notifiers. | | | Although no specific mention made about how the confidential information of the company would be handled by the Indian Government, it is explicitly ensured that no confidential information provided by the individuals to the Government is divulged. |
|---|---|---|---|---|---|

Table 13.1 *Contd.....*

| | | | | | |
|---|---|---|---|---|---|
| 10. | Art. 23(1)(a), 23(1)(b), 23(2), 23(3) | Promote and facilitate public awareness, education and participation concerning the safe transfer, handling and use of LMOs, taking also into account risks to human health. Endeavor to ensure the public awareness and education encompasses access to information on LMOs identified in accordance with the Protocol that may be imported. In accordance with the relevant domestic laws, consult with the public, in decision making under the Protocol, while respecting confidential information. Endeavor to inform the public about the means of public access to Biosafety Cleaning House. | Rule 4(2) & 4(4) | The two central committees namely the RCGM & the GEAC can induct individuals in their capacities as experts to monitor all experiments including research and development on LMOs and products thereof as well as their commercial release. | Public participation and public consultation in the decision-making process regarding use of GMOs and products thereof are part of Indian Rules. |

**Table 13.1** ***Contd.....***

| | | | | | |
|---|---|---|---|---|---|
| 11. | Art. 25(1), 25(2) | Adapt appropriate measures aimed at preventing and, if appropriate, penalizing trans-boundary movements in contravention of domestic measures to implement the Protocol. Dispose, at its expense, LMOs that have been the subject of an illegal transboundary movement through repatriation or destruction, as appropriate, upon request by an affected Party. | Rule 15(1) 15(2) 15(3) 15(4) | These rules deal with penalties if an order is not complied with. | The Rules as well as Art. 15 of the Indian Environment (Protection) Act can initiate enquiry and grant punitive actions on individuals responsible for illegal trans-boundary movements of GMOs & products thereof. |

## Compatibility of Indian legal system with Cartagena Protocol on Biosafety

The Conference of Parties adopted the Cartagena Protocol on Biosafety on January 29, 2000 by more than 130 countries in its extra ordinary meeting held in Montreal, Canada. The Protocol shall be a legal binding instrument to the signatory Member Countries and would come into force after 50 member countries ratified it. It was opened for signature in accordance with its Article 36, from June 5, 2000 to June 4, 2001.The Protocol was signed by the last date, by 103 States, and it entered into force on September 11, 2003, ninety days after the 50$^{th}$ instrument of ratification was received ( as per Article 37 of the Protocol). India ratified the Protocol on January 17, 2003 and its ratification came into force with effect from September 11, 2003. The Conference of Parties (COP) to the Convention of Biological Diversity is the highest decision making body of the Convention. COP held previously 6 meetings before the Biosafety Protocol came into force. COP shall hold its 7$^{th}$ meeting during February 2004 at Kuala Lumpur, Malaysia and shall get merged into the Meeting of Parties (MOP) and would replace the Intergovernmental Committee for the Cartagena Protocol (ICCP) on Biosafety; ICCP was created to prepare the entry into force of the Protocol.

The main requirement for accrediting the protocol by India is to create a nodal agency where there has to be a Biosafety Clearing House and the designated persons for coordinating information exchange and compliance in accordance with the protocol. India has most of the structures in place but it requires building its electronic hardware to create connectivity among the related departments and also with the Biosafety-clearing house of the CBD Secretariat. It also requires intensifying and strengthening its capacity building capabilities in LMOs. It would require considerable sum to create such facilities. The capacity building would require core competence in the evaluation of LMOs as also in the assessment of transgenic traits in any substance that may be suspected to be of LMO origin.

By being party to the Protocol, India shall have a say on the decision making process on the further implementation of the Protocol jointly with other countries. India shall be more visible before the world in contributing to the transboundary movements of LMOs including its own materials. It shall have access to information and expertise from Parties to the Protocol and it shall be eligible to financial support from the Global Environment Facility (which is a financial support receiving mechanism under the Protocol) for strengthening its capacity for handling LMOs. The Indian administrative and legal systems, rules and procedures as well as capabilities in handling LMOs may get broad based. Opportunities and chances of collaborations with other countries are also expected to expand. The decision of India to be a part of the Protocol is therefore considered to be a very right move.

## Indian scenario on transgenic research

### Microorganisms

Efforts are being made to construct transgenic microorganisms that code for bioactive therapeutic proteins. Transformed E. coli coding for several recombinant proteins such as interferons, interleukins, human growth hormone, bovine growth hormone, granulocyte colony stimulating factors, human pro-insulin, human epidermal growth factor, streptokinase, recombinant hepatitis B vaccine etc. are being experimented upon in different laboratories in the country. Transformed E. coli are being currently used for commercial production of granulocyte colony stimulating factor and interferon alpha 2b Recombinant yeast species of Saccharomyces cerevisae, Pichia pastoris and Hanseneulla polymorpha have been modified to code for specific therapeutic proteins. Hepatitis B surface antigen gene has been coded in the above three yeast and the antigen has been isolated and formulated into dosages form for use as vaccines to protect human against Hepatitis B. Such vaccines are currently being commercially made in this country since August 1997. These products have been tested for safety under the Indian EPA and these have been found to be safe. Genetically modified cholera microorganism is being evaluated for its safety as well as efficacy. A non-virulent strain of Vibrio cholera was isolated from natural sources and an immunogenic gene was incorporated into it to make the strain responsible for eliciting immunogenic response and for producing antibodies to it. Genetically modified fungi belonging to Aspergillus spp are being experimented upon for the production of value added, edible enzymes. All such microorganisms are also tested for environmental safety (in case of accidental release in to the open environment) as well as for human safety under the Indian EPA.

## Plants

The first transgenic plant experiment in the field was started in 1995 when *Brassica juncea* plants containing *Bar* gene regulated with plant specific constitutive promoters and linked with *Barnase, & Barstar* genes regulated with floral tissue specific promoters *were* planted at Gurgaon (Haryana), India, under contained conditions. These studies were conducted to assess the extent of pollen escape[6, 7]. Subsequently, several experiments have been started in the field in different locations using transgenic mustard, cotton and tomato. Several Indian institutes and organizations have claimed to develop transgenic plants, which are ready for greenhouse/ poly-house evaluation and some are ready for field evaluation as well. Table 13.2 below gives a list of the major Indian developments up to the present time in transgenic plants.

**Table 13.2 Indian developments in transgenic research and applications**

| Institute | Plants/ crops used for transformation | Transgenes inserted | Aim of the project and progress made |
|---|---|---|---|
| Central Tobacco Research Institute, Rajahmundry | Tobacco | Bt toxin gene Cry1A(b) and Cry1C | To generate plants resistant to *H.armigera* and *S.litura*. One round contained field trial completed. Further evaluation under progress. |
| Bose Institute, Calcutta | Rice | Bt toxin genes | To generate plants resistant to lepidopteran pests. Ready for under-taking Greenhouse testing. |
| Tamilnadu Agricultural Univ., Coimbatore | Rice | Reporter genes like *hph* or *gus A* and GNA gene | To study extent of transformation in the greenhouse. |
| Delhi University, South Campus, New Delhi | Mustard / rape seed | *Bar, Barnase, Barstar* *Marker gene remover(cre-lox)* | Plant transformations completed and ready for greenhouse experiments. Plants with marker genes as well as without marker genes made. |

Table 13.2 *Contd.....*

| Institute | Plants/ crops used for transformation | Transgenes inserted | Aim of the project and progress made |
|---|---|---|---|
| Delhi University, South Campus, New Delhi. | Rice | Selectable marker genes (*hygromycin* resistance and *gus*). Abiotic stress tolerant genes (codA, cor47, hsp1). Cry 1 A c genes | Transformation completed with marker genes as well as with abiotic stress tolerant genes. |
| - do - | Cotton | Abiotic stress tolerant gene (*hva 1*) | Transformation completed. |
| - do - | Wheat | Insect resistance (Pin II) | Transformation completed. |
| - do - | | Abiotic stress tolerant genes (*adc, mtl D, imt I*) | Transformation completed. |
| - do - | Brinjal | Fungal resistance (glucanase) | Transformation completed. |
| - do - | Tomato | CTX-B | Transformation completed. |

Table 13.2 *Contd.....*

| Institute | Plants/ crops used for transformation | Transgenes inserted | Aim of the project and progress made |
|---|---|---|---|
| Indian Agricultural Research Institute sub-station, Shillong | Rice | Bt toxin Gene | To impart lepidopteran resistance, transformations in progress. |
| Central Potato Research Institute, | Potato | Bt toxin Gene | To generate plants resistant to lepidopteran pests. Ready to undertake Simla contained field trials. |
| M/s Proagro PGS (India) Ltd., New Delhi | Brassica/Mustard | *Barstar, Barnase, Bar* | To develop better hybrid cultivars suitable for local conditions; over 15 locations con-ained field trails completed. Further contained open field research trials in progress at multi-locations in 2000-2001. |
| - do - | Tomato | *Cry* 1 Ab | To develop plants resistant to lepidopteran pests; glass house experiments and one season contained field experiment completed Further experiments are in progress. |
| - do - | Brinjal | *Cry* 1 Ab | To develop plants resistant to lepidopteran pests; glass house experiments are in progress. |
| - do - | Cauliflower | *Barnase*, *Barstar* and *Bar* | To develop hybrid cultivars for local use; glass house experiments are in progress. |
| - do - | Cauliflower | *Cry*1H/*Cry* 9C | To develop resistance to pests; glasshouse experiments are in progress |
| - do - | Cabbage | *Cry*1H/*Cry* 9C | To develop resistance to pests; glasshouse experiments are in progress |

**Table 13.2** ***Contd.....***

| Institute | Plants/ crops used for transformation | Transgenes inserted | Aim of the project and progress made |
|---|---|---|---|
| M/s MAHYCO, Mumbai | Cotton | *Cry1Ac* | To develop resistance against lepidopteran pests; Multi-centric field trials in over 51 locations completed and further large scale field trails in progress. |
| M/s Rallis India Ltd., Bangalore | Chilli | Snowdrop (*Galanthus nivalis*) Lectin gene | Resistance against lepidoteran, coleopteran and homopteran pests; transformation experiments are in progress. |
| - do - | Bell pepper | Snowdrop (*Galanthus nivalis*) Lectin gene | Resistance against lepidoteran, coleopteran and homopteran pests; transformation experiments in progress |
| - do - | Tomato | Snowdrop (*Galanthus nivalis*) Lectin gene | Resistance against lepidoteran, coleopteran and homopteran pests; transformation experiments are in progress. |
| Indian Agricultural Research Institute, New Delhi | Brinjal,<br>Tomato,<br>Cauliflower | Bt gene<br>Bt gene<br>Bt gene | To impart lepidopteran pest resistance, transformation completed, green house trials completed and one season field evaluation completed for Brinjal and Tomato. |
| - do - | Mustard/rapeseed | Arabidopsis annexin gene | Transformation completed, Green house trial completed, ready for field trials for moisture stress resistance. |

**Table 13.2** ***Contd.....***

| Institute | Plants/ crops used for transformation | Transgenes inserted | Aim of the project and progress made |
|---|---|---|---|
| I.A.R.I, New Delhi. | Rice | Bt gene | Transformation completed and contained greenhouse trial are in progress. |
| Jawaharlal Nehru University, New Delhi | Potato | Gene expressing for seed protein containing lysine obtained from seeds of *Amaranthus* plants (*Ama-1* gene) | Transformation completed and transgenic potato under evaluation in the contained open environment. |
| Indo Americal Hybrid Seeds, Bangalore | Tomato | Leaf curl virus protein genes, chitinase and alfalfa gluconase gene and combinations of these | Transformation completed, green house tests completed and ready for contained open field experiments. |

## Animals

Work on transgenic animals carried out in India is yet at rudimentary stage. Lab experiments have been conducted to produce transgenic mice containing growth hormone genes. Certain marker genes have also been utilized. By introducing luminescence genes that code for substances that produce light, glowing silkworms were produced; this work only reinstated that silk worms are amenable to the strategy of genetic manipulation. Efforts have also been made to produce transgenic catfish, tilapia and Indian carps containing growth hormone genes; the objectives were to obtain transformants that mature fast. However, these experiments have not provided any significant success, and none of the above work on animals has yet to come to the stage of large- scale trials.

## Conditions for trials using transgenic organisms of all kinds

The RCGM monitors research on transgenic organisms in the laboratory and in the contained open environment/fields. For transgenic plants, experiments are conducted in contained greenhouse to generate several vital safety information before decisions are taken to conduct contained open field experiments. In the field under contained conditions besides designing experiments for collecting data on environmental safety aspects, the agronomic advantages of the transgenic plants in small plots are also assessed. The RCGM looks for information on environmental safety including human and animal food safety issues for all kinds of GMOs. Food safety issues are linked with GMOs that may enter into human or animal food chain directly. The information sought from the trials of GMOs is summarized briefly in Table 13.3 [Ref 5].

**Table 13.3 Summary of the Biosafety information sought from GMO trials**

| Particulars | Information Sought |
|---|---|
| Rationale for the development | Economic, agronomic and other benefits, and rational of development |
| Details of the molecular biology of GMOs | • Description of the host organisms (microorganisms, cell lines, plants, animals etc.)<br>• Source and sequence of transgene<br>• Sequential block diagram of all trans-nucleic acid stretches inserted<br>• Cloning strategy |
| GMOs (Microorganisms) | • Characteristics of expression vectors<br>• Characteristics of inserted genes with details of sequences<br>• Characteristics of promoters<br>• Genetic analysis including copy number of inserts, stability, level of expression of transgenes, biochemistry of expressed gene products etc.<br>• Transformation/cloning methods and propagation strategy. |
| Laboratory, Green House Trials (for plants) and contained enclosure trials (for animals) | • Back-crossing methods for plants<br>• Seed setting characteristics of plants<br>• Germination rates of seeds<br>• Phenotypic characteristics of transgenics<br>• Organism challenge tests where ever applicable<br>• Effects of chemical herbicides for all herbicide resistant plants<br>• Growth characteristics and general health of animals, measured through specific scientific parameters<br>• Toxicity and allergenicity implications to human if any during handling of GMOs |

Table 13.3 *Contd.....*

| Particulars | Information Sought |
|---|---|
| Field trials in open environment | • For GM Plants, comparison of germination rates and phenotypic characteristics, using non-transgenic as controls.<br>• Study of gene flow of plants<br>• Invasiveness studies of plants and animals compared to non-transgenics used as controls<br>• Possibility of transfer of transgenes to near relatives through out crossing / cross-fertilization<br>• Implications of out crossing / cross-fertilization<br>• Comparative evaluation of susceptibility to diseases and pests for plants and animals<br>• For human food / animal feed, elaborate determination of composition and assessment of quality of transformed plants/ fruits/seeds as well as animals as the case may be, with appropriate controls. Compositional analysis shall include near equivalence studies of all the major ingredients in GMOs so as to assess substantial equivalence with reference to non-transgenics. Change in the levels of allergens, toxicants if any, beyond acceptable limits is a matter of food safety concern and such substances are unsuitable for commercial release.<br>• Toxicity and allergenicity implications of transformed GMOs. This include microorganisms, plants/fruits/seeds as well as animals; lab animal studies for food /feed safety evaluation is a requisite.<br>• Handling procedures for allergenic substances.<br>• Agronomic evaluation for GM plants<br>• Economic evaluation for GM animals |

The RCGM authorizes imports of genetic materials; it also permits transfer of such materials within the country. All such movements of transgenic materials are for research use only; decisions are taken based on applications submitted through the IBSC.

For conducting experiments with transgenic plants in contained Greenhouse, designs have been worked out for constructing low cost but substantially contained environment where temperature, light and humidity can be controlled to a considerable extent. Nets have been recommended that arrest the entry/exit of insects below 0.6-mm diameter [Ref 5]. Although similar contained conditions for conducting experiments with transgenic animals have not yet been published, designs are available and can be sent to the investigators on request.

## Issues in Transgenic plant experiments and methods for proceeding step by step

The issues that are taken into consideration before authorizing field trials under contained conditions using GM plants include the potential of the transgenic plants for dissemination into the open environment such as through cross pollination, the dispersal mechanism of the pollens as well as the seeds, the presence of wild members of the species in the eco-system and the presence of other non-transgenic planting materials in the vicinity. While designing field experiments efforts are made to maintain appropriate reproductive isolation so as to prevent the likely-hood of seed setting outside the experimental plot. The transgenic plants are isolated from the gene pool represented by sexually compatible plants to prevent the escape of transgenes. Conditions are also introduced in certain cases to prevent flowering of plants. It is ensured that the genes or the genetically modified plants are not released into the environment beyond the experimental sites. Only such plants are taken into the open environment for experimentation, which have the minimum chance of unintended and uncontrolled adverse affects. The time of sowing, flowering and planting are also taken note of. Only those plants have been used in Indian trials for open field experiments under contained conditions, where the transgenes are considered to be safe or where the pollens are linked with imparting male sterility properties. Experiments have also been designed to study the potential for gene transfer and the consequence of transferring transgenic properties to weeds or other near relatives. The probability of pollen transfer and the natural mutation rates have been made conditions for computation in the experimental designs. The transgenic traits that have been looked at in such experiments in India include Bt-insect resistance, *Bar* resistance, *Bar-barnase* as well as *Bar-barstar* systems, Bar-Bt systems, antibiotic resistance, altered nutritional properties and abiotic stress resistance properties.

As indicated earlier, data for submission by the applicants include mating systems in plants, comparison of germination rate, invasiveness, toxicity and allergenicity or alterations in the anti-nutritional properties of the plants due to the transgenes including the marker genes etc. A detailed format for submitting information has been devised comprising nine chapters, and applicants are required to provide such information to the Government seeking permission for commercial release of target transgenic plants under Rules 7,8,9,10 or 11 of the above Notification [Ref 2].

A few experimental designs have been evolved and approved by the RCGM for conducting trials using GM plants in the open environment. The designs are for studying pollen dispersal, the comparison of cross-ability of non-transgenic plants with the transgenics and evaluation of their comparative competitiveness or invasiveness potential in unmanaged and managed land. The experimental results from two studies had shown that the pollen escape was a real phenomenon. The cross-ability studies conducted for example on transgenic Indian mustard had shown that their existed pre and post fertilization barriers and the results corroborated the classical literature, confirming that escape of transgenes from same crops like the Indian mustard crop was not favoured in nature; however, viable F1 seeds could be produced by manual cross-pollination with related cultivated as well as wild species, which observation was consistent with similar studies made with *Brassica napus.*[Ref 8.] It was observed, while studying the Bt. cotton plants that their pollen also travelled to some distance with the help of insects. It can be stated from these that gene transfer shall be taking place in open environment when transgenic plants are cultivated. By appropriate management practices it might be possible to reduce the extent of pollen transfer into the open environment for all crops, but it cannot be fully contained. Therefore, the consequence of gene transfer is an issue, which is real. It is to be ascertained whether the country can take advantage of the technology by keeping a close watch on the effects of pollen escape and renew its decisions from time to time by periodically reviewing the ground realities. The implications of this issue have not yet been satisfactorily resolved. A decision has to be taken by the Indian Government on this. The experience of other countries can set the ground for a strategy on this.

In addition to these experiments, major chunks of data are required to be generated on food safety in accordance with the latest Indian guidelines[5.] The guidelines emphasize quantitative production of transgenic proteins and their effects on as-is-where-is basis on experimental animals in the context of determining the toxicity allergenicity and anti-nutritional properties etc. The data generated in Indian experiments for Bt. cotton at Industrial Toxicology Research Center, Lucknow using goat as the ruminant model, and for transgenic Indian mustard assessed on rat, rabbit, guinea pig and hen model (at Shriram Institute of Industrial Research, Delhi) as well as on goats model (at Fredrick Institute of Plant Protection and Toxicology, Tamilnadu) did not show any additional food safety risks.

The transgenic field experiments conducted in India have enabled the country to have hands-on experience on several genetically modified plants. Most important among them are transgenic Bt cotton, *Bar-Barnase* and *Bar-Barstar* mustard and Bt tomato. Data generated in India has demonstrated substantial agronomic benefits from transgenic plants over the corresponding non-transgenic controls. Table 13.4 provides an overview of the initial findings on the performance of the GMO plants.

**Table 13.4 Percentage increase in productivity of transgenics over non-transgenics**

| Name of the plant | Range of increase in productivity in % over controls | |
|---|---|---|
| Bt Cotton | : 23 to 60% | : Average in 51-location study was nearly 40% |
| Bar-Barnase-Barstar mustard | : 5% to 43% | : Average in 15 location study was over 16% |
| Bt tomato | : about 300% | : (one location study) |

Several other plants such as brinjal, cabbage, cauliflower, rice, pigeon pea, soybean, chilli, bell pepper and corn have been transformed with improved traits and these are soon likely to be experimented upon in open environmental conditions.

Based on exhaustive environmental safety studies as well as food safety evaluation, the Government of India permitted the cultivation of insect resistant Bt cotton in March 2002 in three tetraploid hybrid lines, designated as MECH-12, 162 and 184. The field performance of these lines had been satisfactory in fields where the necessary inputs in terms of water, fertilizers and pesticides (for sucking pests) were applied; in regions where there were droughts or inadequate supply of water and other inputs, the results were not as were anticipated. Gaining experience from 2002, it is anticipated that the future years shall witness better results by deploying this technology with appropriate inputs.

To summarize, the Indian Government created the rules and procedures for dealing with GMOs in 1989 under the Indian EPA. Several recombinant DNA products have been approved for production and marketing in India. Presently there are four production units for recombinant DNA drugs. Three rDNA drugs are being made in India presently but eleven have been approved for marketing, the remaining once are imported and used. As time elapses, more units would be coming into operations and rDNA drugs would be made available at a very competitive prices to the consumers.

In GM plants, the first transgenic plant experiment in the field was started in 1996. Since then, the country has acquired substantial experience in understanding the issues related to the handling of GMOs. The rich experience gained from hands on experiments in field conditions has provided insights about the benefits as well as the risks from the use of such plants. On overall assessment it appears that the benefits from the use of such plants are substantial and they may outweigh the risks, which in many facets are presumptive and not real. With regard to actual risks, there are issues on the flow of transgenes into the open environment, which are indeed real. Transgenic traits would spread into the natural environment over the years. Such spread of traits has not yet been found to create environmental problems that cannot be contained. The country has to decide on how it would weigh the agronomic benefits vis-à-vis the issue of transgenic pollen escape into the environment. The author has reviewed the progress of transgenic plant experiments in India [Ref 9-11]. As stated above, the introduction of Bt cotton by the regulatory authorities in 2002, and its field results subsequently found, are gradually showing that the technology is, indeed, powerful. It is expected that by adapting to this technology, India would soon be able to increase cotton productivity and reduce cultivation costs to become a stronger player in international cotton textile trade. Many other transgenic technologies for application in Indian agriculture are in the offing.

Experiments in transgenic animals including in fish are yet at the developmental stage and the country has to go a long way before such products are developed for commercial applications.

After the COP-MOP meeting on Biosafety in Kuala Lumpur, Malaysia in February 2004, it is anticipated that the transboundary movements of LMOs may get accelerated. India would have to take steps to comply with the requirements of the Protocol. India already has good infrastructure and experience in place, which would need to be strengthened and streamlined. It would be easier for India to comply with the Protocol and take advantage from it. Towards these directions, more efforts must therefore be initiated.

### *References*


1. Indian Environment (Protection) Act, 1986.
2. Rules for the manufacture, use, import, export and storage of hazardous micro organisms/genetically engineered organisms or cells, issued by the Union Ministry of Environment and Forests, Government of India, vide Notification No. G.S.R. 1037 (E) dated 5$^{th}$ December 1989.
3. Recombinant DNA Safety Guidelines, 1990 issued by the Department of Biotechnology, Union Ministry of Science and Technology, Government of India in 1990.
4. Revised Guidelines for Safety in Biotechnology, issued by the Department of Biotechnology, Union Ministry of Science and Technology, Government of India in May 1994.
5. Revised Guidelines for Research in Transgenic Plants & Guidelines for Toxicity and Allergenicity Evaluation of Transgenic Seeds, Plants and Plant Parts, issued by the Department of Biotechnology, Union Ministry of Science and Technology, Government of India in August 1998 and partly amended on 24. 9.1999.

6. Ghosh, P.K., *Current Science,* 1997, 72(3), 172-179
7. Ghosh, P.K., *J. Natl. Bot. Soc.,* 1997, 51, 11-32.
8. Schaeffler, J. A. and Dale, P. J., *Transgenic Res.,* 1994, 3, 263-278.
9. Ghosh, P.K., and Ramanaiah T.V., *J.Sci.&Ind.Res.*, 2000, 59, 114-120
10. Ghosh, P.K., *Genetically modified crops in India with special reference to cotton,* Silver Jubilee Lecture Series, Lecture delivered on June 23, 2000 at Indian Society for Cotton Improvement, C/o Central Institute for Research on Cotton Technology, Mumbai – 400 019, 1-26.
11. Ghosh, P.K., *The Rules for Handling Transgenic Organisms And Biopharmaceuticals Produced Therefrom in India,* Journal of Sci. & Ind. Res., Vol. 62,June 2003,554-572.

# 14

# Current Status of Transgenic Crops in India

*** C. R. Bhatia**

A number of laboratories are engaged in developing transgenics in different crops to enhance insects, diseases, abiotic stresses and improving nutritional qualities. In contrast, the private seed companies have imported the transgenic stocks developed by their business partners abroad, and used accelerated back crossing to transfer the genes into locally adapted genotypes. Regulations for research, field experiments, and approval for commercial cultivation by the farmers are in place. Three cotton hybrids developed after back crossing to "Bollgard" Bt-cotton were approved for commercial cultivation in 2002. About 90,000 small farmers grew Bt-cotton in one acre plots. Science based studies show that the pesticide application for cotton bollworm control was reduced, and the

* Former Secretory, Dept of Biotechnology, Government of India; D17 Rohini, Sector 9 A, Vashi, New Bombay - 400 703.

net income of the farmers increased. The report sale of seed in 2003 is reported to be good. As in other parts of the world, there are opponents of tranasgenic crops in India. They link transgenics with other social, and economic issues such as globalization, privatization, WTO, control of seed industry by multinational corporations, and intellectual property rights. Among the various risks from the transgenic crops, the gene flow of the transgenics to other cultivars, landraces, wild and weedy relatives has emerged as a real risk. This could be a major concern with herbicide resistant transgenics. The report made by the Science Academies, including the Indian National Science Academy, the Brazilian Academy of Sciences, and the Third World Academy of Sciences represents well- balanced view-point. Major concerns for the small farmers in developing countries have been addressed in this report.

There has been considerable excitement, and interest among the Indian plant scientists for standardizing transformation – regeneration protocols for different crop plants since the first reports of transfer, and expression of bacterial gene(s) into tobacco in 1983. Our laboratory at the Bhabha Atomic Research Center, Bombay was the first to report transfer, expression, and inheritance of *npt2* gene into model plant tobacco in 1987[Ref 1]. This was essentially the repetition of the experiments first reported from US and Europe, but was necessary for our capacity building. This was followed by transfer of *bar* gene for herbicide resistance into tobacco[Ref 2]. transformation of *Brassica juncea*[Ref 3]. indigenous cloning of *Bacillus thuringiensis* entomotoxic protein gene [Ref4]. and its transfer and expression in tobacco [Ref 5].

As many laboratories in the country had strong traditions of plant tissue culture, and reasonable infrastructure, they initiated transformation – regeneration experiments using the imported constructs. The Department of Biotechnology supported five Centers for Plant Molecular Biology in existing laboratories. They initiated gene cloning, and development of new constructs. At about the same time generous support under the International

Rice Biotechnology Network funded by the Rockefeller Foundation contributed to rapid developments in transgenic research[Ref 6]. Nevertheless, the first set of cotton hybrids with *Bacillus thuringiensis* gene were approved for commercial cultivation in the year 2002.

## Biosafety regulations

Rules for the manufacture, use, import, export and storage of hazardous microorganisms / genetically engineered organisms / cells and plants were framed under the Environment Protection Act in 1989 by the Ministry of Environment and Forests [Ref 7]. The rDNA safety guidelines were issued in 1989 [Ref 8] by the Department of Biotechnology that has the nodal responsibility for rDNA research. These were based on the recommendations of the Recombinant DNA Advisory Committee taking note of similar guidelines in USA and European countries developed earlier.

Three level system of approvals is followed, and the boundaries of each committee are specified. First the Institutional Biosafety Committee (IBSC) of each of the laboratories approves and oversees the transgenic research of their scientist. At the second level is the Review Committee on Genetic Manipulation (RCGM) which permits and oversees contained as well as small plot (one acre) field experiments with transgenic crops. The design and layout of the field experiments is approved by RCGM. It also appoints Monitoring and Evaluation Committees (MEC) to visit each of the field experimental sites and report to RCGM. The biosafety guidelines were revised in 1994[Ref 9], and more detailed guidelines for transgenic crops were issued in 1999[Ref 10]. At the third level Genetic Engineering Approval Committee (GEAC) functions under the Ministry of Environment and Forests. Large scale field experiments, and final approval for commercial production are authorized by the GEAC. This is based on the data submitted by the party seeking approval, MEC and GEAC. Besides the above, there are State Biotechnology Coordination Committees for each state, and District level Committees (DLC) in each district (county), mainly to ensure implementation of the biosafety guidelines.

While some people feel that the clearance procedure is too complicated, others hold the view that the present evaluation system for biosafety is not adequate. Newspaper articles for and against the decisions of the GEAC follow. It has been argued that the GEAC should approve the gene from the biosafety considerations, and not only the specific gene-crop genotype. Others insist on complete biosafety evaluation even when the same gene is transferred to different genotypes by back crossing. While the Bt cotton was under consideration, the regulators insisted on carrying out animal feeding experiments in the country using the local Bt cotton. This gene (*cryIAc*) was the one developed by Monsanto, and already used extensively in USA, Australia and China. The Australian regulating agency accepted the feeding data generated in USA, and did not insist on repeating the experiments locally. However, the regulators in India wanted that the animal feeding experiments should be repeated in the country.

## Current status of transgenic crops

There are many projects in the Indian laboratories to develop transgenic crops. While publicly funded laboratories are indigenously developing transgenic lines, the private seed companies have imported the transformed lines from their business partners abroad, and have followed accelerated back crossing program to incorporate the desired genes into locally adapted cultivars/hybrids. The crops are cotton, rice, potato, tobacco, tomato, *Brassica* species, egg plant (*Solanum melongena*), banana, pigeonpea (*Cajanus cajan*), chickpea (*Cicer arietinum*) blackgram (*Vigna radiata*), peanut (*Arachis hypogaea*) and castor bean (*Ricinus communis*). The details can be found in a special section in Current Science brought out this year[Ref 11].

Table 14.1 gives the transgenic plant material imported from abroad mainly by the seed companies. They are developing locally adapted cultivars following back crossing program. Few approvals for commercial cultivation are expected from these. Program for transferring the nutritional traits of "Golden Rice" developed in Switzerland has been initiated. Table 14.1 lists the ongoing research efforts on developing transgenic rice (*Oryza sativa*).

**Table 14.1. Transgenic planting material imported in India**

| Crop/cultivar | Transgene | Source | Indentor |
|---|---|---|---|
| ***Brassica juncea*** | *barnase* | Belgium | SC. |
| | *bar,barstar* | Belgium | SC. |
| | *barnase, bar* | Belgium | SC. |
| | *barnase, barstar, bar* | Belgium | SC |
| ***Brassica napus*** | *osmads 1, bar* | Belgium | SC. |
| ***Brassica oleracea*** var. capitata | Cry 9C | Belgium | SC. |
| ***Cicer arietinum*** | *Bean Alpha A1* | Scotland | ICRI |
| | *Bean Alpha A1* | Australia | U |
| ***Glycine max*** | *Cp4epsps, aad* | USA | SC. |
| | *Cp4epsps, aad* | USA | SC. |
| ***Gossypium sp*"** "Bollgard" | *CrylAc* | USA | SC. |
| | *Cp4epsps, aad* | USA | SC. |
| | *cryX* | USA | SC |
| | *vip3* | USA | SC |
| ***Oryza sativa*** | *Ama1* | Philippines | U |
| | Herbicide Basta resistance | USA | PRI. |
| | *Cry1Ac* | UK | SC |
| | *CryAB, cry19C, bar* | Belgium | SC |
| | *Cry1AB, Xa21, PR* | Philippines | ICRI. |
| | Phytoene synthase, Phytoene desaturase, Lycopene cyclase | Switzerland | PRI*. |
| | *Xa21* | Philippines | SC. |
| | *Cry1Ab, bar* | Belgium | SC. |
| ***Zea mays*** | *Cry1Ab* | USA | SC. |
| | *Cp4epsp0s* | South Africa | SC. |
| | *Cry1Ab, pat* | USA | SC |

Information based on [26]

ICRI = International crop research institute, PRI = Publicly funded research institute, SC = Private seed company, U = University.

Stocks imported by the seed companies are being used in accelerated back crossing programs to incorporate the transgenic into local cultivars / hybrids.

* Transgenics Golden Rice.

**Table 14.2 Current research programs for rice transgenics in India**

| **Insect resistance** |
|---|
| *Bacillus thuringiensis* genes *cry1Ab* and *cry1Ac* and their fusion products |
| *cry 9Aa*, synthetic *cry1Ac* |
| Proteinase inhibitor, which inhibits gut proteinases of several Lepidopteran insects |
| Potato protease inhibitor 2 (*Pin-II*) |
| Garlic leaf lectin – *Allium sativum* agglutinin (ASAL) |
| Snow drop (*Gallianthus nivalis*) lectin – GNA gene |
| **Disease resistance** |
| *Xa21* gene from *Oryza longistaminata* for bacterial blight resistance |
| Thaumetin like protein (*tlp*) |
| b 1, 3-glucanase gene |
| Rice chiinase (*chit 1*) gene |
| *Trichoderma* chitinase gene |
| *Arabidopsis* NPR 1 gene which regulates salicylic acid mediated systemic acquired resistance |
| **Virus resistance** |
| Rive tungro bacilliform virus (RTBV) promoter |
| Rice tungro spherical virus (RTSV) coat protein |
| **Stress resistance** |
| Barley late embryogenesis abundant protein (Lea 3) |
| Moth bean pyroline-5-carboxylate synthesis (*PSCS-129-A*) gene |
| Transit peptide plus manitol1-phosphate dehdrogenase |
| (*TP-mt1DH*) gene |
| Choline oxidase (*codA*) gene from *Arthobacter globiformis* for glycine-betaine |
| Accumulation associated with stress tolerance |
| Inositol synthase gene from wild halophytic rice *Porteresia coarctata* |

**Table 14.2** ***Contd....***

Genes involved in polyamine biosynthesis

Mouse ornithine decarboxylase (*odc*) in sense and antisense

Oats arginine decarboxylase (*adc*) gene

Human adenosyl methinine decarboxylase (*samdc*)

*Datura* spermidine synthase (*spd syn*)

Annexin gene of *Arabidopsis* in sense and antisense

**Male sterility**

*T-urf-13* – a maize gene associated with Texas cytoplasmic male sterility

RASP – rice anther specific promoter

**Multiple genes**

Five minimal transgenes cassette hygromycin phosphotransferase (*hpf*), phosphinothricin transferase (*bar*), β-glucuronidase (*gusA*), luciferase (*luc-1*), and anthrinlate synthesis (*as*) genes

From [22]. Information based on Abstracts of the 8th National Rice Biotechnology Network Meeting held during 21-25 October 2002 at Aurangabad, organized by MAHYCO Research Foundation, Mumbai.

Considerable efforts are towards development of plant-incorporated protectants (PIPs)[Ref 12] in cotton, rice, *Solanum melongena*, pigeon pea, and chickpea. The most successful program supported by public funds has been the development of transgenic cotton transformed with indigenously designed Bt. ∂-endotoxin by Rakesh Tuli and his associates at the National Botanical Research Institute, Lucknow. *cry1EC* and *cry1Aa* have been designed for high level expression in plant cells, synthesized and used for transforming cotton[Ref 13]. The technology has been transferred recently to a consortium of small seed companies who propose to transfer the two genes into their cotton hybrids[Ref 14]. Other significant developments are the transfer of *ama1* gene from *Amaranthus* to potato to enhance the nutritional value of the latter by Asis Dutta and his colleagues at Jawaharlal Nehru University, New Delhi. *OXDC* gene for oxalate de carboxylase from *Amaranthus* has been transferred to tomato for fungal disease resistance.

In March 2002 three transgenic Bt hybrids developed by MAHYCO were approved for commercial cultivation in India. They were grown over an area of 95000 acres. The crop faced severe drought conditions in many areas. The data collected by the scientists of the University of Agricultural Science, Dharwad[Ref 15] shows that cultivation of Bt-cotton resulted in reduction of pesticide use for bollworm control. While 4-9 insecticide sprays were required for the NHH-44 hybrid, NHH-44 with Bt required only 3-4 applications. The cost of cultivation was reduced. Net increase in profit was 42 per cent (Rs. 13,521; 1 US$ = Rs. 45). However, some NGO's have reported no advantage of Bt-cotton, and net loss to the farmers due to higher cost of seed. The repeat sale of seed is reported to be good for the planting season of 2003.

## Clandestine cultivation of Bt-cotton

While the official approval for Bt-cotton was being debated, extensive cultivation of cotton resistant to bollworm was reported from the state of Gujarat. The crop of farmers who had obtained the seeds from a particular seed company was almost free from bollworm while the neighboring plots were heavily damaged. Further tests established that this crop did carry a Bt gene. The seed company pleaded ignorance, stating that they were not aware that the stock they selected was transgenic. This was a clear violation under the environmental protection act. There was demand for the destruction of the crop that was nearing harvest. This was not acceptable to the farmers, and the State Government was not able to implement the decision. The proprietor of the seed company was arrested, and subsequently released. The legal battle continues in the court. Meanwhile, this clandestine Bt-cotton has been grabbed by many farmers and has spread to other states.

## Cost of developing and commercializing transgenic crops

In the editorial of the Current Science special section on transgenic crops the guest editor raised the question "How is it that while in the laboratories a large number of transgenics have been developed, very few are in the field." In this context, this author[Ref 16] has pointed out that the overall cost of developing and commercializing transgenic crop cultivar / hybrid involves the following major components – (i) transfer of the desired gene into the crop species with adequate expression, (ii) development of crop cultivar / hybrid combining other desirable agronomic and quality parameters, (iii) nutritional, toxicological and other biosafety evaluation in laboratory and field experiments to meet the requirements of the regulatory process, (iv) agronomic evaluation in field experiments and (v) seed production and distribution. Most of the laboratories engaged in developing transgenics are well organized for step (i) only. The development cost of a marketable transgenic crop cultivar / hybrid is nearly equivalent to that of bringing a new drug to the international market[Ref 17] which is currently estimated at US $ 800 million. The cost may be somewhat lower in the publicly funded research institutions in the developing countries, but would still be substantial. This makes it extremely difficult for publicly funded scientists supported through project grants to generate data for the regulatory agencies. Considering the need of transgenic cultivars for enhancing productivity and sustainability, evolving partnerships between public research institutions and private seed companies, or NGOs is essential in the developing countries.

## Environmental impact of transgenic crops

A large number of environmental concerns with respect to the transgenic crops, first raised by Rissler and Mellon[Ref 18] are repeated again and again by the opponents in spite of the fact that the adverse effects have not been observed. Bhatia and Mitra[Ref 19] had examined the same in the Indian context. The global results show that the early concerns mainly reflected the "fear of the unknown", and most of the anticipated adverse effects have not been observed. However, the possibilities of transgenic crops out crossing with other crop cultivars of modern agriculture, land races, "traditional" populations, and wild, or weedy relatives of the crop plants, and consequent transgene flow have emerged as a reality based on scientific data [Ref 20-23]. Out crossing is ubiquitous in both self and cross pollinated species. Crossing between sympatric populations and gene flow between them at low probabilities is well known. Transgenics have revived the interest, and funding in this area. Gene flow was addressed in two international conferences recently. It is currently one of the major research areas targeted by the Biotechnology Risk Assessment Research Grants Program of the United States Department of Agriculture. Measures to block or substantially reduce hybridization in order to decrease the environmental impact are being developed. Insect and disease resistant transgenics are not expected to have any significant effect, as the wild species often show higher levels of resistance. Multiple herbicide resistance in volunteer *Brassica napus* has been attributed to gene flow from different herbicide resistant *B. napus* cultivars[Ref 24]. Gene flow and its possible consequences will be highly region specific, depending on the presence of sympatric species, and need to be examined on case-to-case basis and considering the area of cultivation.

## Opposition to the transgenics

The local activists have been extensively supported, and fed with mis-information by similar groups in Europe. Transgenic crops have been muddled with other social and political issues, such as privatization, globalization, World Trade Organization, protection of intellectual property rights (patents) and their effect on product prices, multinational companies (MNCs), MNCs control of seed industry, neo-colonization etc. These issues attract newspaper headlines, and public attention, while the scientific facts are forgotten. They are also not presented properly, and do not make "news". Few years back, the opponents entirely misrepresented the facts, and linked the so-called "terminator" technology (which was still at the experimental stage) with the transgenics and made the farmers and politicians to believe that the harvested seeds will not germinate. It has caused considerable damage to the transgenics.

In this context, the report of the Science Academies[Ref 25] represents well-balanced view-point. Many of the provisions included therein were incorporated through the interventions of the Third World Academy of Sciences. The interests of the developing world that need the new technologies more than the developed, food surplus countries ware taken care off. Collaborations among the developing countries can also help in accelerating the pace.

## The road ahead for developing countries

In order to feed its growing population, India has no other alternative but to increase the productivity of the existing good farmland. Otherwise population pressures will lead to further destruction of forests, and cultivation of marginal lands. This is also true for most other developing countries. Transgenic technology enlarges the gene pool available for improving the crop plants. The benefits of the first generation transgenics already available far outweigh the possible risks that are manageable.

### *References*

1. Viegas, P., H. Mathews, C. R. Bhatia and N. K. Notani (1987), Monohybrid and dihybrid segregation's in the progeneis of tobacco transformed for Kanamycin resistance with Ti vector system. J. Genetics, 66: 25-31.

2. Viegas, P. and N. K. Notani (1993), Heritable transformation of tobacco by Agrobacterium mediated transfer of Streptomyces derived herbicide resistance gene bar. J. Genetics, 72: 35-42.5.

3. Mathews, H., N. Bharathan, R.E. Litz, K.R. Narayanan, P.S. Rao and C.R. Bhatia (1990), Transgenic plants of mustard *Brassica juncea* L.Czern and Coss.Plant Science,72: 245-252.

4. Tuli, R., J. Saluja and N. K. Notani (1989), Cloning and expression in *Escherichia coli* of entomotoxic protein gene from *Bacillus thuringiensis* subspecies krustaki. J. Genetics, 68: 147-160.

5. Tuli, R. et al. (1997), Development of transgenic plants for resistance to insect attack. In, Plant Molecular Biology and Biotechnology (Eds. K. K. Tiwari and G. S. Singhal) Narosa Publishing House, New Delhi 37-42.

6. Herdt, R. W. Establishing the Rockefeller Foundation's priorities for rice biotechnology research in 1995 and beyond. In, IRRI International Rice Research Institute. 1996. Rice Genetics III. Proceedings of the Third International Rice Genetics Symposium, 16-20 Oct.1995. Manila (Philippines) IRRI.

7. Rules for the manufacture, use, import, export and storage of hazardous microorganisms / genetically engineered organisms or cells. 1989. Ministry of Environment and Forests, Govt. of India. Notification No. GSR 1037 (9E) of December 5, 1989.

8. Recombinant DNA Safety Guidelines (1990). Reprinted July 2002, Department of Biotechnology, Govt. of India, New Delhi. p 89.

9. Revised Guidelines for Safety in Biotechnology. (1994)., Department of Biotechnology, Govt. of India, New Delhi.

10. Revised Guidelines for Research in Transgenic Plants and Guidelines for Toxicity and Allergenicity Evaluation of Transgenic Seeds, Plants and Plant Parts 1998. Department of Biotechnology, Govt. of India, New Delhi. p. 92.

11. Special Section: Transgenic Crops (2003). Current Science, 84: 297-424.

12. Tuli, R., C. R. Bhatia, P. K. Singh, and R. Chaturvedi (2002)., Release of insecticidal transgenic crops and gap areas in developing approaches for more durable resistance. Current Science, 79: 163-169.

13. Singh, P. K., M. Kumar, C. P. Chaturvedi, D. Yadav, S. Ashraf, S. V. Savant and R. Tuli (2003). , Development of transgenic cotton for resistance to insect pests: Comparison of three indigenously designed delta endotoxin – coding genes targeted for protection against *Helicoverpa armigera* and *Spdoptera litura*. Current Science (in press).

14. Seed cos tie up to buy *swadeshi* Bt cotton tech. Economic Times, Bombay, October 15, 2003.

15. Indian study confirm superiority of Bt-cotton in fighting pests. Business Standard October 9, 2003.

16. Bhatia, C. R. (2003), Cost of developing and commercializing transgenic crops. Current science, 84: 1499-1500

17. The Economist, 27 March 2003, p. 15.

18. Rissler, R. and M. Mellon (1993). Perils Among the Promise – Ecological Risks of Transgenic Crops in a Global Market, Union of Concerned Scientists, Cambridge, Massachusetts pp 92.

19. Bhatia, C. R. and R. Mitra (1998). Biosafety of transgenic crop plants. Proc. Indian National Science Acad., Part B, 64: 293-318.

20. Gene Flow from GM to non-GM Populations in Crops, Forestry, Animals and fishery Sectors, FAO Electronic Forum in Biotechnology in Food and Agriculture. http://www.fao.org/biotech; e-mail:biotech–room 3@mailserv.fao.org.

21. Scientific Methods Workshop on Ecological and Agronomic Consequences of Gene Flow from Transgenic Crops to Wild Relatives. The Ohio State University, Ohio, 5-6 March, 2002.

22. Bhatia, C. R. and R. Mitra (2003), Consequences of gene flow from genetically engineered crops. Current Science, 84: 138-141.

23. Willkinson, M. J. et al. (2003), Hybridization between *Brassica napus* and *B. rapa* on a national scale in the United Kingdom. Science, 302- 457-459.

24. Hall. L., K. Topinka, J. Huffman, L. Davis and A. Good (2000). Pollen flow between herbicide-resistant *Brassica napus* is the cause of multiple resistant *B. napus* volunteers. Weed Science 48: 688-694

25. Transgenic Plants and World Agriculture., Report prepared under the auspices of the Royal Society of London, the USA National Academy of Sciences, the Brazilian Academy of Sciences, the Chinese Academy of sciences, the Indian National Academy of Sciences, the Mexican Academy of Sciences and the Third World Academy of Sciences, 2001, The Royal Society, p 18.

26. Mangal, M., K. Malik, and G. J. Randhawa (2003), Import of transgenic material: National Scenario. Current Science, 85: 454-58.

# 15

# Pre and Post Release Monitoring of GMOs

* J.S. Bentur

Genetically modified organism (GMO) is a plant, animal or micro-organism produced by genetic engineering in which foreign genes encoding desirable traits are transferred from one organism to another across barriers of sexual incompatibility. Advances in molecular biology during last two decades have helped to develop technologies for creation of GMOs and opened up a new domain of commercial exploitation of the technology. As a result, a range of GMOs and biotech products is now available and is being expanded with increasing speed. Realizing the inherent benefits and potential risks, concern is being expressed, since the advent of technology, by scientists, environmentalists and general public about

* Principal Scientist (Entomology) Directorate of Rice Research, Hyderabad - 500 030, jbenture@yahoo.com

'biosafetyness' of GMOs. Some of the related issues are presented here mainly in the context of genetically modified crop plants with special emphasis on transgenic *Bt* crops.

## Risks of genetic engineering

There may not be generic harms associated with GMOs like in case of radiation or first generation pesticides. All genetically engineered crops may not produce toxic food or not all the genes incorporated into them may escape in the environment. But specific engineered organisms may be harmful by virtue of the novel gene combinations they possess. Thus risks of genetically engineered organisms must be assessed case by case and that these risks can differ greatly from one gene-organism combination to another. So far, scientists have identified a number of ways in which genetically engineered organisms could potentially adversely impact both human health and the environment. Once the potential harms are identified, the question will be known how likely are they to occur. The answer to this question falls into the arena of risk assessment. In addition to the known potential there may also be risks that we simply do not know enough to identify. The recognition of this possibility does not by itself justify stopping the technology, but does put a substantial burden on those who wish to go forward to demonstrate the benefits.

## Potential harms to human health

Here are some examples of the potential adverse effects of genetically engineered organisms may have on human health. Most of these examples are associated with the growth and consumption of genetically engineered crops. Different risks would be associated with genetically engineered animals/microorganisms and, like the risks associated with plants, would depend largely on the new traits introduced into the organism.

***New allergens in the food supply :*** Transgenic crops could bring new allergens into foods that sensitive individuals would not know to avoid. Genetic engineering routinely moves proteins into the food supply from organisms that have never been consumed as foods. Some of those proteins could be food allergens, since virtually all known food allergens are proteins. The problem is unique to genetic engineering because it alone can transfer proteins across species boundaries into completely unrelated organisms. An example is, transferring the gene for one of the many allergenic proteins found in milk into vegetables like carrots. Soybeans genetically engineered to contain Brazilnut proteins have been shown to cause reactions in individuals allergic to Brazil nuts. Some farm workers exposed to *B. thuringiensis* pesticide were shown to have developed skin sensitization and IgE antibodies to the *Bt* spore extract. *Bt* toxin *Cry1Ac* has recently been shown to be a potent oral/nasal antigen and adjuvant . There is no way of predicting this except through experience.

It may be noted that human race has evolved through constant exposure to a variety of allergens such as those contained in pollen, dust and a variety of foods. Most of the allergenic proteins range in molecular weight 10 –70 kDa and are generally resistant to breakdown by heat and acids. Several of the normal food materials are known to contain harmful ingredients like kidney bean with toxic lectin, cassava and peach seeds with cyanogenic glycosides. Milk, egg and sea foods contain proteins that are allergenic to some sensitive people.

***Antibiotic resistance :*** As of date, most of the plant genetic engineering protocols use antibiotic resistance genes as "selectable markers". Early in the transformation process these markers help select cells that have taken up foreign genes of our interest. Although they have no further use, the genes continue to be expressed in plant tissues. These antibiotic-resistance genes produce enzymes that can degrade antibiotics taken with meals. If a tomato with an antibiotic-resistance gene is eaten at the same time as an antibiotic, it could destroy the antibiotic in the stomach. Besides, the resistance genes could be transferred to human or animal pathogens, making them impervious to antibiotics. If transfer were to occur, it could aggravate the already serious health problem of antibiotic-resistant disease organisms.

Although unmediated transfers of genetic material from plants to bacteria are highly unlikely, any possibility that they may occur requires careful scrutiny in the light of seriousness of antibiotic resistance. Further, in most cases the antibiotic used for selection in plant transformation is kanamycin which is not used in humans and used to a very limited extent in animals. Alternative selection systems are being developed which do not involve antibiotic resistance genes.

***Production of toxins :*** Crop plants normally produce toxins to fend themselves from pathogens and herbivores. However, often plants contain inactive pathways that lead to production of toxic substances, which gets activated once the pathogen attacks. Addition of new genetic material through genetic engineering could untimely activate these inactive pathways leading to increased and sustained high levels of toxic substances within the plants.

***Concentration of toxic metals :*** Some of the new genes being added to crops can remove heavy metals like mercury from the soil and concentrate them in the plant tissue. The purpose of creating such crops is to make possible the use of municipal sludge as fertilizer. Sludge contains useful plant nutrients, but often cannot be used as fertilizer because it is contaminated with toxic heavy metals. The idea is to engineer plants to remove and sequester those metals in inedible parts of plants. In a tomato, for example, the metals would be sequestered in the roots; in potatoes in the leaves. Turning on the genes in only some parts of the plant requires the use of genetic on/off switches that turn on only in specific tissues, like leaves. Such products pose risks of contaminating foods with high levels of toxic metals if the on/off switches are not completely turned off in edible tissues. There are also environmental risks associated with the handling and disposal of the metal-contaminated parts of plants after harvesting.

***Enhancement of the environment for toxic fungi :*** Although for the most part health risks are the result of the genetic material newly added to organisms, it is also possible for the removal of genes and gene products to cause problems. For example, genetic engineering might be used to produce decaffeinated coffee beans by deleting or turning off genes associated with caffeine production. But caffeine helps protect coffee beans against fungi. Beans that are unable to produce caffeine might be coated with fungi, which can produce toxins. Fungal toxins, such as aflatoxin, are potent human toxins that can remain active through processes of food preparation.

***Unknown harms :*** As with any new technology, the full set of risks associated with genetic engineering have almost certainly not been identified. The ability to imagine what might go wrong with a technology is limited by the currently incomplete understanding of physiology, genetics, and nutrition.

## Potential Harms to Environment

***Gene transfer to wild or weedy relatives :*** Novel genes placed in crops will not necessarily stay in agricultural fields. If relatives of the altered crops are growing near the field, the new gene can easily move via pollen into those plants. The new traits might confer on wild or weedy relatives of crop plants the ability to thrive in unwanted places, making them super weeds. For example, a gene conferring weedicide resistance or insect resistance may provide with added advantage to the weed. This in turn might prove a challenge for the crop production system.

***Change in herbicide use patterns :*** Crops genetically engineered to be resistant to specific herbicides are tightly linked to the use of particular chemical pesticides. For instance cultivation of Glyphosate Tolerant Soybean (GTS) farmers need necessarily use glyphosate for weed control. Adoption of these crops could therefore lead to increased use of the specific herbicides changing the present mosaic mix of chemical herbicides used across the country. This changing pattern could result in greater levels of environmental harm. In addition, this could lead to the rapid evolution of resistance to the herbicides in weeds, either as a result of increased exposure to the herbicide or as a result of the transfer of the herbicide trait to weedy relatives of crops. Again, since herbicides differ in their environmental harm, loss of some herbicides may be detrimental to the environment.

***Squandering of valuable pest susceptibility genes :*** Many insects contain genes that render them susceptible to pesticides. Often these susceptibility genes predominate in natural populations of insects. These genes are a valuable natural resource because they allow pesticides to remain effective. The more benign the pesticide, the more valuable the genes that make pests susceptible to it. Certain genetically engineered crops threaten the continued susceptibility of pests to one of nature's most valuable pesticides: the *Bacillus thuringiensis* or *Bt* toxins. These "Bt crops" are genetically engineered to contain gene(s) for the Bt toxin. More details on *Bt* crops are dealt in a separate section below.

***Toxicity to non-target organisms :*** Addition of foreign genes to plants could also have serious consequences for wildlife in a number of circumstances. For example, engineering crop plants for insect resistance might lead to toxic effect on predators and parasites through food chain. Expression of toxins like protease inhibitors might affect cattle, which feed on the crop residues. Frog, fish and birds that feed on target insects may also be affected.

***Creation of new or worse viruses :*** One of the most common applications of genetic engineering is the production of virus-tolerant crops. Such crops are produced by engineering components of viruses into the plant genomes. For reasons not well understood, plants producing viral components on their own are resistant to subsequent infection by those viruses. Such plants, however, pose other risks of creating new or worse viruses through two mechanisms: recombination and transcapsidation. Recombination can occur between the plant-produced viral genes and closely related genes of incoming viruses. Such recombination may produce viruses that can infect a wider range of hosts or that may be more virulent than the parent viruses. Transcapsidation involves the encapsulation of the genetic material of one virus by the plant-produced viral proteins. Such hybrid viruses could transfer viral genetic material to a new host plant that it could not otherwise infect. Except in rare circumstances, this would be a one-time-only effect, because the viral genetic material carries no genes for the foreign proteins within which it was encapsulated and would not be able to produce a second generation of hybrid viruses.

***Unknown harms :*** As with human health risks, it is unlikely that all potential harms to the environment have been identified. Each of the potential harms above is an answer to the question, "Well, what might go wrong?" The answer to that question depends on how well scientists understand the organism and the environment into which it is released. At this point, biology and ecology are too poorly understood to be certain that question has been answered comprehensively.

## Risk assessment

Having identified a list of possible harms that might occur as a result of using or releasing genetically engineered organisms, the next question is how likely are any of these to occur? Like the original "brainstorming" of potential harms, the answer to this question depends greatly on how well the organisms and their interaction in the environment are understood.

Risks must be assessed case by case as new applications of genetic engineering are introduced. In some circumstances, it is possible to reduce or remove the risks with great confidence. But for many potential harms, the answers are far less certain. Risk assessments can be complicated. Because even rigorous assessments involve numerous assumptions and judgment calls, they are often controversial when they are used to support particular government decisions.

## Pre-release monitoring

Pre-release testing for the presence of allergins and toxins in the GMOs and safety tests are mandatory in developed countries before the product is submitted for approval for extensive field testing. However, such tests are often subject to severe criticism (Pusztai, 2001) on the ground that the tests are poorly designed, results not subject to rigorous statistical analysis and not peer reviewed while the products are passed on as "substantially equivalent". It is concluded that in the absence of reliable methods for allergenicity testing, it is at present impossible to definitely establish whether a new GMO is allergenic or not before its release into the human/animal food/feed chain.

Pre-release evaluation of long-term ecological risks can be limited to estimation of the probabilities that an initially rare transgene can spread into the ecosystem. Spread of the transgene into natural populations may result in a number of ways, including (i) vertical gene transfer as a result of matings with wild relatives, (ii) invasion of new territories through introduction and (iii) horizontal gene transfer mediated by microbial agents, or a combination of these factors. The relative importance of each factor is dependent on species, transgene inserted, and method used to insert the transgene. Inter-breeding between crop plants and their wild relatives is a common and natural phenomenon observed for several crops including rice, maize and cotton. Indeed, this is one of the pathways breeders have adopted to improve crops through traditional breeding approach to confer pest resistance and abiotic stress tolerance. It is also shown that there is enhanced propensity of out crossing in transgenic plants. However, whether transfer of a weedicide resistance or insect resistance gene alone will make a wild species to be a super weed needs careful assessment. Probability of squandering of valuable pest susceptibility gene is high. Following case study will illustrate the situation and focus on ways to mitigate the problem.

## Case study of *Bt* crops with insect resistance

*Bt* crops have been a large commercial success. Farmers in USA were the first in 1996 to grow *Bt* maize, cotton and potato. During the year 2002, of the 58.7 million ha. covered by transgenic crops in 7 industrial and 9 developing countries, *Bt* crops accounted for about 14.7 million ha. or 25 percent of the total (James, 2002). Economic advantage gained during 1999 by *Bt* cotton alone is estimated to be $ 213 million.

These *Bt* cultivars contain genes from the ubiquitous bacterium *Bacillus thuringiensis* (*Bt*) and encode insecticidal proteins known as delta-endotoxins. These are highly toxic to certain insect pests but are generally safe to humans. Genetic engineering with *Bt* genes is a powerful technology for protecting crop against some kind of insects. Even before the advent of biotechnology, *Bt* was used as natural pesticide in crude and purified form as spray formulation against pests of horticultural crops and forest trees. Despite the limitations like stability and uniformity, *Bt* pesticide was filling the void created by loss of effectiveness of several of the chemical pesticides following development of resistance in target insects. Insects have evolved resistance to all classes of widely used insecticides, including *Bt* products. Because the *Bt* crops produce the toxin in most plant tissues throughout the life cycle of the plant, pests are constantly exposed to it. This continuous exposure selects for the rare resistance genes in the pest population and in time will render the *Bt* pesticide useless. Realizing the strength and weakness of *Bt* transgenic technology the Union of Concerned Scientists had warned the developmental and regulatory agencies in USA on the dire need of imposing a set of "resistance management" regulations for the *Bt* crops on 'now or never' basis.

Over the past 10 years, scientists, government regulators, environmentalists and the private sector have vigorously debated if and how the evolution of pest resistance to *Bt* crops can be delayed. There is wide spread agreement on one strategy – the "high dose/refuge" (see Cohen *et al.*, 2000). This strategy is being enforced for *Bt* cotton, maize and potato in the USA. In 1997 a sub-committee for Insect Resistance Management (IRM) was constituted to evaluate and address scientific issues related to risk assessment and formulate strategies for resistance management. Pre-release monitoring as envisaged in the regulation is to verify if the transgenic *Bt* lines identified for clearance meet the requirement of high dose with specific target pests. Additionally, pre-release monitoring and testing of target pest populations for their susceptibility and establishing baseline toxicity data not only help in identifying effective toxins but also identify geographic limitations for the proposed transgenic line. It is also desirable that transgenic crops may contain more than one unrelated *Bt* genes or a *Bt* and non-*Bt* insecticidal gene in combination to extend the life of the two genes together. The IRM sub-committee has also helped in development of options for maintenance of refuge areas based on expert committees. These expert committees have developed guidelines based on the current information on biology, ecology and genetics relating to the pest, crop plant and production systems in different regions. For instance, refuge recommendations for *Bt* corn are based on decision trees developed specifically for the purpose. Based on the perceived risk level like (i) negligible or low, (ii) medium or (iii) high, refuge area of 5, 10 or 40 per cent of total area if refuges are not to be protected or 10, 20 and 80 per cent if refuges are to be protected was recommended for the year 1999 for the Northern USA.

## Post-release monitoring

With awareness of the specific perceived risks associated with cultivation of *Bt* crops, objectives of post-release monitoring are clearly defined. Monitoring of compliance by growers of regulations relating to refuge maintenance and protection practice is primary. Scouting for evidences of buildup of tolerance to *Bt* toxin in target and non-target pest populations based on field monitoring and lab testing is critical. Other research inputs are also taken into consideration while periodically revising and updating IRM recommendations. Effective maintenance of refuge area that is grown to non-*Bt* crop as per specifications is regulated through licensing system in USA.

During 1998, following the reports that *Bt* corn pollen can harm Monarch butterfly population and green lacewing predators feeding on larvae from *Bt* corn, had higher mortality than those feeding on larvae from non-Bt corn, a second ecological risk concerning cultivation of *Bt* corn has been added. Accordingly, EPA modified conditions for *Bt* corn for the year 2000. These required planting 20 percent refuge with insecticide application only in areas where the pest population exceeds economic thresholds. Following the indications that tolerance to *Bt* is growing among bollworm populations in cotton over last three years, a change in refuge pattern in cotton for the year 2001 was suggested. Growers could choose one of three structural refuge options : (i) 5 percent external refuge, at least 15 feet wide within ½ mile linear distance, not to be treated with any insecticide labeled for the control of tobacco budworm, cotton bollworm, or pink bollworm, (ii) 20 percent external refuge within 1 mile distance and all crops to be treated by insecticide except foliar *B.t.k.* products or (iii) 5 percent internal refuge within the *Bt* field and all crops to be treated by insecticide except foliar *B.t.k.* products. A community refuge program was allowed as a pilot for the 2001 growing season wherein all farmers could participate and identify the refuge area collectively. Thus resistance management strategy is kept dynamic to be improved over time. *Bt* corn and cotton are now being grown for 5 years and there are no reports of development of substantial resistance in pest populations so far. This may be attributed to the fore thought of scientists and implementing agencies and growers cooperation in observing the regulations.

## Biosafety regulations in the Indian context

From the foregoing account it is clear that for successful harnessing of biotechnology for sustainable agriculture importance of biosafety regulations cannot be over emphasized. Though biosafety regulating bodies are identified under the aegis of Department of Biotechnology, these need to be more visible to the general public and reports generated by the experts need to be made public for scrutiny and criticism as being practiced in the developed countries. *Bt* crops are the first released for cultivation. But it is not certain if we have the concomitant Resistance Management Strategy elaborately worked out. A committee, if any, formed should have drawn a panel of experts with wide expertise and experience to formulate the guidelines. Even baseline toxicity data for the specific toxins for the Indian populations of the target pests are not available. Besides, a strong regulating agency is also needed to oversee compliance by growers on maintenance of refuges and appropriate monitoring of pest populations.

Participation of well-informed citizens, scientists from different fields and general public is yet another ingredient for the success. We need to have voluntary, non-profit organizations serving as media for information, inviting public debates, scientific and critical independent evaluation of the reports and to guide policy makers.

*References*

Cohen, M. B., F. Gould and J.S. Bentur*. 2000. *Bt* rice: Practical steps to sustainable use. International Rice Research Notes 25:4-10. (*for reprint in pdf contact the author)

James, C. 2002. Global Review of Commercialized Transgenic Crops: 2002. ISAAA Briefs No. 29

**(http://www.isaaa.org/press**_Rerelease/Briefs29_2003/Gmreview_Nov6.htm).

Pusztai, A. 2001. Genetically modified foods – are they a risk to human/animal health? **(http://www.biotech-info.net/pusztai_article.html)**

**Relevant web sites :**

**www.biotech-infor.net**
**www.ucsusa.org**
**www.isaaa.org**

# 16

# Biotechnology and Construction of Safety Parameters

* E. Haribabu

Biotechnology has implications for production in agriculture, horticulture, animal husbandry and medicine. Genetic engineering, the most important of the biotechnological interventions, has become controversial because of its implications for risk, safety and sustainable development.

Technology, as a body of techniques and the knowledge underlying the techniques, is the medium through which human-beings transform nature. In the process of transformation artifacts are generated that have both use value and exchange value. History of technology shows that over time, especially with the capitalist mode of production, the emphasis has been on producing artifacts that have exchange value in the market. Risk and

* Professor, Department of Sociology, University of Hyderabad - 500 046

safety are two important dimensions of any technology that is developed. Earlier risk analysis has been confined to economic domain. Further, with the increasing awareness about the environmental consequences of technologies, analysis of risk for the environment has become one of the central questions. Similarly benefit-cost analysis, which was earlier concerned with economic benefits and costs, now has to reckon with social costs. For example, the questions, who pays the cost and who get benefits is related to issues of justice and equity. Relevant questions here are : What are the differentials in risk-bearing capacity among different social groups? How do the differentials account for the distribution of risk?

## Risk and Safety Analysis in a Sociological Perspective

A risk is an uncertain potential harm to human life or health or to property or environment (Barbour 1980 p. 175). When options have multiple objectives and multiple consequences, the separation of means and ends is more difficult. Analysis of potential risk associated with biotechnology may be contrasted with that of the risks and benefits associated with green revolution. Earlier, risk analysis made distinctions in terms of technical and economic dimensions. However, the social dimension was ignored. Risk analysis should be carried out by incorporating the social dimension as well. One can argue that what is technical is also socially arrived at. Pinstrup-Anderson and Marc Cohen (1999) point out three essential differences between green revolution and biotechnology, which have a bearing on the risk and benefits. One, the research efforts that led to green revolution were concentrated in public sector institutions, the intellectual property rights did not extend beyond the initial release. Once the farmers acquired the seed they could reuse it without further payment although the reused hybrid seed reduced the yield. In contrast modern agricultural biotechnology research is undertaken by the private sector firms, which protect IPRs through patents beyond the first release. Farmers cannot reuse the seed without the permission of the patent holder for the next season. A second related difference between the Green revolution and the Gene revolution involves the patenting of processes as well as products that indicates a shift towards proprietary research process. The third difference is that while the Green revolution technology was focused on the specific problems of the developing countries. The Gene revolution focuses on research that is suitable for industrial country. The transfer of agricultural biotechnology from the industrial countries would ignore the specificities of the developing countries and the transfer will not be an adaptation of technology but direct transfer. For example, Bt crops have been developed for North American cropping environment. The crop varieties containing Bt gene require extremely knowledge – intensive cultivation.

There are two important aspects to risk analysis: (a) judgments on acceptable levels and (b) time element.

## Judgments on Acceptable Levels of Risk

Judgement of 'acceptable levels' of risk are never purely scientific when weighing of incommensurable costs and benefits involves trade-offs among diverse values (Barbour 1980 p. 127). The differentials in the evaluation of risk by different groups and the degree of affordability of risk are important sociological variables. What is acceptable risk by the yardstick of the industry may not be acceptable by farmers, different risk-bearing capacities also influences what level of risk is acceptable. For example, what is perceived as a risk by an Indian farmer holding 2 acres may not be seen as a risk by an Indian farmer holding 10 acres or a North American farmer holding a few thousand acres. Policy-relevant studies and advice can never be value-free, even when carried out by scientists and engineers (Skolnikoff and Harvey Brooks 1975).

## Time Element

What is the time frame over which it is safe and less risky to use a particular technology? How long the Bt cotton will confer resistance? Can we give the approximate time frame? In terms of environment what are the consequences over time, say next generation? Answer to these questions have to be generated on the basis of experiments/projections/simulation studies. For example, consider the idea of refuge in which conventional variety is grown along with a genetically engineered crop to avoid risks associated with insects becoming resistant, say, the Bt gene. How much land has to be devoted to conventional varieties to work as refuge? Is it 5 per cent or 10 per cent or 20 per cent of the land under Bt seed? Can a small farmer who owns 2 ha of land afford to leave 20 per cent of the land as refuge? If Bt gene can be transferred to a target organ in the plant, which part of the Bt plant may be made refuge? Consider another strategy: missing Bt seed and conventional varieties and planting them in the same plot so that a separate refuge is not necessary. Then the question is what are the proportions of the Bt seed and conventional seed that have to be used? These options have to be evaluated for their safety in a given social context.

## GMOs and Safety Issues

Genetic modification of organisms, through deliberate human intervention, is an invention of a different kind compared to inventions based on inorganic matter-gadgets, machinery, and other equipment.

The nature, number of safety parameters that are important and the levels of tolerance have to be constructed. As outcomes of biotechnology research and development are closely related to the interests of several groups-industry, scientists and technologists, government, farming communities, and consumers-issues relating to identifying safety parameters, their definition and scope and fixing limits of tolerance can not be left to any single group with its narrow interests. Dommelen (1999) suggests that a rigorous set of relevant questions (SRQs) which he defines as a collection of research questions that a scientist considers relevant for the study of a research problem have to be formulated to identify issues related to hazards during pre-release and post-release phases. The social construction of the safety parameters have to include question arising from a broad based of interested parties, not just a narrow group with self-interests in approval for genetically engineered organism (Dommelen 1999).

## Relevant Questions in a Sociological Perspective

Questions seen as relevant by one group may not be seen as relevant by another group from its point of view. Over the last one decade the industrialized countries of Europe and North America have generated enormous amount of material on safety issues in the context of genetic engineering research and its application. Any question that we raise about safety and potential risk in the Indian context will have to be from the perspective of Indian cropping environment and resource – endowments of the highly differentiated farming communities in India.

## Pre-release R&D Phase

The precautionary principle that one should not undertake activities about which there is scientific uncertainity about their impact should guide the discourse on safety of biotechnology. Safety concerns of the scientific workforce involved in the biotechnology R & D during development phase arise out of: (a) uncertainity associated with outcome of their research at the level of cognition; and (b) laboratory environment in which they handle several hazardous materials including radio-active substances. In the case of research connected with genetic modification of the organisms – plants and animals – as it is difficult to foresee the consequences of the genetic modification for the environment and there should be adequate containment facilities. In India, in most of the public sector institutions involved in genetic engineering research, the containment facilities are grossly inadequate. As the R&D scientists are quite knowledgeable about the properties of the materials they use in their work environment. They tend to handle materials carefully with adequate knowledge of the consequences of their negligence. The managements that control the laboratories, both in the **private sector and the public sector** generally put in the place laboratory safety norms so that the managements do not have to face legal problems in the event of failure of safety measures. The following institutional mechanisms deal with biosafety issues in India.

## Biosafety regulatory mechanisms in India.

Biosafety regulations were first prepared in 1990 and revised in 1994. the regulations aim to cover the research as well as the field trials of GMOs both locally developed and imported GMOs. The regulations are implemented by the following committees at different levels.

1. Institutional Biosafety Committee (IBSC) is mandatory for every institution engaged in research activities involving genetic manipulation of microorganisms, plants and animals, to set up a ISBC. The ISBC includes representative from the concerned institution, the Department of Biotechnology (DBT) and a doctor.

2. Review Committee on Genetic Manipulating (RCGM) is a committee under the DBT and its mandate is to monitor biosafety implications of all ongoing research.

3. Genetic Engineering Approval Committee (GEAC), The Ministry of Environment, through its GEAC, is responsible for biosafety in the cases of field trials and large scale release of GMOs. The committee also monitor production, sale and import, use of genetically engineered products and import, export, transport, handling and use of the GMOs.

## Risk and Safety in post-release phase and farming community

Farming community in India, as mentioned above, is highly differentiated in terms of resource-endowments and access to technologies, credit and information. As mentioned above, cultivation of genetically modified crops is knowledge-intensive that is basically developed for cultivating environments in industrialized countries where the operational holdings are quite large ranging from a few hundred hectares to thousands of hectares. As of now farmers do not have institutional protection against potential risks and safety failures backed by legal support. Each farmer has to fed for himself/herself in an event of risk. Individual farmers may have to fight legal battles against big and powerful seed companies. The idea that as producer, farmers have rights over what they produce - the food grain and the seed is very much a part of the farmer's worldview.

From the point of view of farming community the relevant questions are: Are the inputs economically affordable ? Is the technology resource-neutral ? Is it compatible with their meaning systems (culture) ? Is it safe and less risky ? Can we generate answers to these questions on the basis of experiments/simulation studies ? How do we address the question of equitable access to technologies-products, information regarding management and credit ?

### Industry

The private initiative occupies a dominant place in the biotechnology R&D. The primary motive in the present context of the propriety process of research the primary motive of the private corporations would be related to patents, profits and market expansion. Some of the important questions are: Will the industry give the whole truth? There is a justifiable perception that industry believer in maintaining brevity with truth. Does the industry provide complete information regarding package that goes with GM seed technology? Example: questions relating to maintaining refuge as mentioned above, whether or not insecticide and pesticide sprays are required or not, if they are required to what extent they are required? To what extent the claims on the genetically modified food are made on the basis of an objective and transparent evaluation?

## Risk and Safety in the Consumers' worldviews

Consumers of genetically modified food/fruits/medicine tend to evaluate the products from the point of view of their meanings, derived from culture, that they attach to what they consume as food. We have to understand the perceptions and meanings. Whether or not GM food is safe to consume from the point of view beliefs associated with health and nutrition is an important question.

Whether or not GM food is compatible with meaning, system associated with religion and environment is also an important issue.

What are the short term and long-term risks involved in consuming GM foods?

Answer to these questions have to be generated on the basis of research.

How do we resolve the knotty questions? One way of resolving these questions is through process of negotiation that involves all the stakeholders. Negotiation between parties that are unequal in terms of power relations may result in compromise rather than consensus.

### *References*

Barbour, Jan G. 1980 Technology Environment and Human Values, New York : Praeger.

Dommelen, Ad van. 1999 Hazard identification of Agricultural Biotechnology: Finding Relevant Questions, Utrecht, The Netherlands: International Books.

E. Haribabu, 2002 Cognitive Empathy in inter-disciplinary research: the contrasting attitudes of plant breeders and molecular biologists towards rice. Journal of Biosciences (J. Biosci), vol.25, No.4, pp 323-330.,

Pinstup-Andersen per and Cohen Marc J. 1999 'Modern Biotechnology for Food Agriculture: Risk and Opportunities for the Poor' in G.J. Presiey and M.M. Lantin (eds) Proceeding of the International Conference on Biotechnology, conference convened by the CGIAR and the US National Academy of Science.

Skolnikoff and Harvey Brooks. 1975 'Science Advice in the white house?' Science, 197, 1975).

# 17

# Intelllectual Property Right In Agricultural Biotechnology Implications For Entrepreneurs And Farmers

* **B. Venkateswarlu**

The issue of intellectual property rights has come into centre stage ever since the signing of WTO agreement. TRIPS (Trade Related Intellectual Property Right) is an important component of the WTO agreement. Though it is not part of the text of the Agreement on Agriculture (AOA), it has tremendous impact on the AOA as it directly influence the generation and transfer of new agricultural technologies. Historically, developed countries have strong intellectual property rights. The IPR's have served as the spring board for the growth of industrial development in the west. IPR's are also protected in India but it is considered to be weak protection. The Indian Patent Act, 1970 protects only the process patents. After signing of WTO,

* Principal Scientist, Central Research Institute for Dryland Agriculture, Santhoshnagar, Hyd - 500 059

it has become necessary that the IPR protection in India matches with other countries in the world. This chapter briefly reviews the concept of patenting, current status of IPRs in India, changes made in Patent Act to comply with the WTO agreement and its likely impact on scientific research, small farmers and entrepreneurs.

## Concept and forms of intellectual property right

Intellectual property is basically a creation of human mind. It is the starting point for development of any industry, product or service. People all over the world realized quite early that protection of intellectual property right encourages creativity and innovation. This protection when translated into a legal right, often results in economic benefits to the individual besides benefiting the society at large. All along, IPRs were strongly supported in most western countries while in the east, knowledge is not considered as an individual's property and hence there was no protection or weak protection.

There are different forms of intellectual property rights. These are:

- Patents
- Copy rights & related rights
- Trademarks
- Geographical Indications
- Industrial Designs
- Layout Designs for Integrated Circuits
- Undisclosed Information

In addition, database right and traditional resource rights are also being considered to be included in the ambit of IPRs. Each form of IPRs has its own significance for a particular industry or trade, but patents are considered to be the most important forms of IPRs.

## What is a Patent ?

A Patent is an exclusive right to use or exercise an invention granted to a person for a limited period of time in consideration of the disclosure of the invention. The patent is granted by the state to an individual or group of persons. It must be remembered that a patent is granted to a person because it involves invention, which only a human mind can do. However a person can always assign the patent to an organization in which he is employed. The following are the important requirements for patenting.

- Novelty
- Non-obviousness (inventive step)
- Industrial use (practical use)

The novelty of an invention is determined by search of the prior art. Non-obviousness means that the invention should not be an obviously known fact to the person skilled in the art. It should involve an inventive step and not a mere discovery of already existing thing. Not only inventions related to new products and processes can be patented, but also new uses of already known products qualify for patenting. The following are further important attributes of a patent i.e.

- A patent has a time limit (15-20 years)
- A patent has a geographical limit
- A patent is restricted to the claims made in the application

However every invention, howsoever new it might be, cannot be patented; some are excluded, which are listed below:

- Inventions contrary to order public or morality, including that protect human, animal or plant life or health or to avoid serious prejudice to the environment.
- Diagnostic, therapeutic and surgical methods for the treatments of human or animals
- Plants and animals other than microorganisms, and essentially biological processes for the production of plant or animals other than non-biological processes.

Countries, however, shall provide for the protection of plant varieties either by patents or by an effective sui generis system or a combination thereof.

## How to file a patent

The inventor has to file a patent in the concerned regional office in the country in case of national patents. With regard to international patents, the guidelines and requirements of the country in which patenting is proposed must be thoroughly understood and then proceeded. The procedures, fee and the duration required etc., vary from country to country. Publication in the gazette and inviting opposition to the patent from public also differ between countries. A flow chart of patent application and its sealing (approval) in India is given below with approximate time taken for each step.

**Filing a Patent in India – The Flow Chart**

Submit an application form to the relevant patent office

↓ 12 months

Preliminary examination by the patent office after filing complete specifications

↓ Effective date

First examination report (Accept/abandon)

↓ 15 Months

Second examination report (Accept/abandon)

↓ 4 Months

Opposition from public

↓ 2 months

Sealing of patent and registration

## Patents in agriculture

Historically agriculture was excluded from patenting in most countries. However some countries allowed protection of plant varieties in the form of "breeder's rights". Over the years, this protection has gradually increased and some countries like USA started allowing patents also, if there is an inventive step. Generally new plant varieties world over are protected under a set of guidelines under the umbrella of UPOV (International Union for the Protection of New Varieties of Plants). The UPOV legislation was modified from time to time from 1961 to 1991. During this process it became more in favour of breeders and seed companies. Some countries like USA offer protection to seeds, bulbs and fish etc., both under UPOV and patents. However inputs used in agriculture like pesticides, harmonies and farm machinery and tools are protected under patents. The salient differences between the plant variety rights and patents are shown in the following Table 17.1.

**Table 17.1 Comparison of different provisions under plant variety right and patents**

| Provision | Plant variety right (Under UPOV 1991) | Patent law |
|---|---|---|
| Protection coverage | Plant varieties of all genera and species | Inventions |
| Requirements | Novelty, distinctness, Uniformity, stability | Novelty, Inventiveness, Non-obviousness |
| Protection term | Minimum 20 years | 17-20 years (OECD) |
| Protection scope | Commercial use of all material of the variety | Commercial use of protected matter |
| Research exemption | Yes | Variable |
| Breeders exemption (commercial use) | Not for essentially derived varieties | No |
| Farmers privilege | Up to national laws | Limited form in prospect in EU countries |

**Source :** Michael Blakeney, Joel I. Cohen and Stephen Crespi, 1999 - Intellectual Property Rights and Agricultural Biotechnology in Managing Agriculture Biotechnology – Addressing Research Program Needs and Policy Implications (ed. J.I. Cohen) CAB International, pp 209-227.

In India "Method of agriculture and horticulture" is totally excluded from patenting in the Indian patent Act, 1970, but process patents in agro-chemicals and agro-procesing are allowed. After signing of WTO agreement, the Indian patent Act has been amended to comply with the provisions of the TRIPS. The salient features of the new patent Act are described in the following section.

## WTO agreement and its impact on IPRs in India

For the first time in the history of GATT, agriculture was also included in the agreement at the Uruguay round in 1986. The draft agreement on agriculture which was part of the Dunkel draft was debated for its positive and negative features in all countries. TRIPS provisions as contained in the final agreement made it mandatory for all countries to enact suitable legislations to extend the protections of IPR's in agriculture also as shown in the following specific clauses.

- Article 27 makes it mandatory for all countries to protect inventions in all fields without discrimination of field of technology.
- 27.3 (b) provides some relief to plants and animals (other than microorganisms) i.e., protection could be by (a) Patents (b) sui generis system or (c) both.

However adequate flexibility was allowed to member countries in drafting the legislation appropriate for their countries keeping in view the nature of agriculture and the level of protection needed. Accordingly, the Government of India has introduced a bill entitled–Plant variety protection and farmers rights Act, 2001 (PVPFR) which was passed by Lok Sabha but awaiting the approval of Rajya Sabha.

The salient features of the PVPFR Act–2001 are described below :

- Recognizes farmer's contribution and confers him specific rights.
- Right to save, resow, exchange or sell seed in the same manner as he was doing earlier.
- Right of compensation for under performance.
- Burden of proof not on farmers (unlike TRIPS 34)
- Recognition of community rights.
- Setting up of gene fund
- Compulsory licensing to avoid monopoly and profit making
- Protection of extinct and farmers varieties

In many ways, the Indian PVPFR Act-2001 is considered revolutionary in its content. It recognizes specific farmers rights which allows them to exchange seeds (including GM seeds) in the same manner as he was doing in the past. It may be noted that in India 85% of the seed requirement is met by farmer to farmer exchange. However he cannot sell the seed with a brand name. Another unique feature is providing a right to farmer to register a variety in his/her name just like the breeder. In this case, the farmer need to just satisfy the DUS (Distinctiveness, Uniformity, Stability) criteria instead of NDUS (Novelty, Distinctiveness, Uniformity, Stability) as required in case of breeders. This provision is likely to give a fillip to the conservation effort of local land races by the farmers and community organizations.

Another related Act passed by the parliament is the Geographical indication of goods Act-1999 which protects the unique names of our commodities originating in a particular geographical region of the country because of which they became popular and have a premium in the trade.

Basmati rice, Darjeeling tea, Alfonaso mangoes and Nagpur oranges are some of the best examples of geographical indications. With the passing of this act, it is expected that we will be able to protect some of our well known agricultural commodities from being patented or trade marked by companies in other countries.

## Salient features of the new patent act

To comply with the WTO, the Indian Patent Act (second amendment-1999) has introduced many new provisions and significantly changed some existing ones. With regard to agriculture, the new Act offer relatively more protection to the researchers.

The recent advances in the field of Biotechnology have also been considered in making these amendments. While the basic tenets and philosophy remain more or less same with regard to patenting of plants and animals, the amended Act provides some good incentives for inventors in the field of biotechnology and genetic engineering. The comparision of the existing IP Act. 1970 and the amended Act is shown in the following Table 17.2.

**Table 17.2 Comparison of different provisions related to agriculture and biotechnology under the IP act, 1970 and the amended act, 2002**

| Item | Under the IP Act, 1970 | Under the Amended Act 2002 |
|---|---|---|
| Life forms | Not Patentable | Patentable |
| Plants and animals | Not Patentable | Not Patentable |
| Microorganisms | Not Patentable | Patentable |
| Microbial and non-biological processes | Petentable | Patentable |
| Biochemcial and Biotechnological processes | Non Patentable | Patentable |
| Plant varieties | Non Patentable | Protection under PVFP |

Table 17.2 *Contd...*

| Item | Under the IP Act, 1970 | Under the Amended Act 2002 |
|---|---|---|
| Method of Agriculture and Horticulture | Not Patentable | Some process qualifying for novelty patentable |
| Tissue culture, cells, viruses, cell lines, vectors, and plasmids | Not Patentable | Not patentable |
| Gene therapy | Not Patentable | Not patentable |
| Process of producing useful substances, MC antibodies, enzymes etc. using GMO's (Hybridoma technology) | Not Patentable | Patentable |
| r-DNA Technology | NA | Not Patentable |
| Transgenic plants and animals | NA | Not Patentable |

**Source:** Shanti Kumar, 2001, Patentability of biotechnology innovations. Paper presented at the workshop on *Intellectual Property Rights in Indian Agriculture – An Awareness, NAARM, Hyderabad- May 31, 2001.*

## Impact of IPR's on biotechnology research and new varietal development

With the emergence of the biotechnology as a powerful tool for development of genetically modified crops and varieties, new concerns have arisen on the impact of TRIPS in technology development and transfer between rich and poor countries. In the past, agricultural research was mostly in the public domain and technology know-how and materials moved freely between countries. But with biotechnology playing a significant role in varietal development, the initiative has now gone into the private sector as this type of research requires large capital expenditure. Moreover, development of a particular product involves permission to use patented processes and material from several sources, involves significant amounts of royalty payments. This increases the cost of the final product. New inventions are protected immediately by private corporations, which are not easily available to public sector research institutions. In many developed countries, patenting is allowed for gene sequences, promoters, vectors, transformation system, diagnostic probes and markets. Every year, hundreds of applications are made for patent protection in these areas in the American and European patent offices. While this competitive spirit promotes new innovations, it also restricts the availability of technology to scientific community at large.

For any private sector company, return on investment is an important criteria and therefore protected technologies are used either for their own product development or licensing to other research partners at high premium. This will only add to the cost of the final product. Some research foundations like Rockfellers and Ford have facilitated a certain degree of public funding for biotechnology research by supporting international research networks, but these are not enough. Collaboration between universities and research institutions in the public sector with private seed companies could be one solution in developing countries to face this challenge. The pro's and con's of strong intellectual property rights in the Indian context are given below.

## Pros

- Strong IPRs promote competitive sprit in creativity and innovations.
- Strong IPRs attract investments from private sector into R&D
- Third largest scientific manpower in India which will have incentive to invent
- The whole society benefits from new technologies in the long run

## Cons

- The prices of patented products can go up
- May lead to monopolies in some advanced fields
- Companies (and MNCs) with financial muscle may dominate the scene
- Objective of self-reliance may be adversely affected

The developed countries which already have strong IPRs are moving towards stronger protection leading to the development of what is known as "proprietary" science. Multinational companies (MNCs) with high R&D budgets are likely to edge out smaller local companies with small budgets. The GM seeds which are likely to come over of this kind of high capital intensive research and the associated biosafety costs are likely to be more expensive than the traditional hybrids. There is also fear that such new technology may be affordable at least in the initial stages only to the resource rich farmers. Therefore the Government of India and the State governments have to step in to subsidize such new technologies among the millions of small and marginal farmers as was done in the beginning of the green revolution for HYV seeds. While the new technologies might benefit agriculture in general in the long run, there are bound to be certain difficulties during the transition phase for the developing countries on one hand and small and marginal farmers on the other.

## Biodiversity and IPR

India is one of the seven-mega biodiversity centers. We have extensive traditional and indigenous knowledge on plants, animals, traditional medicine etc., India is home for an estimated 15000 plant species, 67000 Insects, 4000 Mollusks, 61500 Invertebrates, 1200 Birds and 340 Mammals. Such extensive genetic diversity offers both immediate and long term economic opportunities. However, with the entry of IPR's into plants and animals, urgent steps are needed to document and catalog all the relevant information on our biodiversity in order to prevent others taking undue advantage. For example due to lack of patent awareness and proper documentary evidence on traditional uses of our biodiversity, as many as sixty patents on uses or products from neem, thirty on tamarind, nineteen on turmeric, two on black pepper, two on cucumber, two on karela, two hundred and fifty on medicinal plants, two on ginger and two on gymnema were taken by US and European companies. The biodiversity bill, 2000 will hopefully address the issues of bio-piracy effectively. In any case, we need to realize that the future advances in medicine and agriculture would largely depend on the wise use of our plant and animal biodiversity.

## Recommendations

From the foregoing account, it is obvious that certain immediate steps are needed to protect our biodiversity, encourage innovation/patenting and evolve suitable policies so that the domestic industry, small entrepreneurs and farmers are not adversely effected. Besides the Government, which can evolve policies and suitable legislative measures, non-governmental organizations have also to play a major role in educating the public and the farmers. Following are some specific recommendations.

- Increased public sector spending in frontier areas like agri-biotechnology for developing "our own" technologies
- Increase awareness in the R&D institutions on IPR protection
- Strong public versus private sector collaboration
- Education of farmers, NGOs and other stakeholders on the value of biodiversity, provisions of the PVPFR and Gene fund.
- Greater vigilance on movement of biological material across borders
- Preparation of biodiversity registers, documentation of indigenous knowledge and early listing of geographical indications.

### *References*

Lecture notes on patents, document 022, Technology Information, Forecasting & Assessment Council (TIFAC), Department of Science & Technology, February, 2001.

Bala Ravi.S. TRIPS and Protection of Plant Varieties and Farmers Rights Paper presented at National Workshop on "Challenges of Liberalization to Indian Agriculture and Protection of Plant Varieties and Farmers Rights", NIRD, Hyderabad, 6-7 July, 2001.

Joel. L. Cohen, Managing Agricultural Biotechnology- Addressing Research Program Needs and Policy Implications, CABI International.

# 18

# Genetic Engineering in Crop Improvement

* P. B. Kirti

Crop Biotechnology has become a reality with transgenic crops occupying an estimated 58.7 million hectares world-wide, an area more than twice that of Great Britain. Emphasis was initially on crops like soybean, maize, cotton and canola for controlling weeds and insects. Now, other crops and a wider range of traits are being targeted.

## Conventional Crop Improvement

Conventional plant improvement is based on the following steps

1. Generation (through mutagenesis)/introduction (through successful hybridization between selected genotype) of genetic variability.

* Professor, Department of Plant Sciences, University of Hyderabad, Hyderabad - 500 046.

2. Selection for superior types from the generated variability
3. Multiplication of selected superior types
4. Testing the selected genotypes in different environments (multi-location testing)
5. Release of a tested genotype as a variety

Of the above steps, the crucial ones are generation of variability through physical/or chemical mutagenesis and introduction from other plants through hybridization. Though the field of mutagensis dominated the scene in the 1970s the progress in crop improvement was not significant probably because of indiscriminate use of the technique. The conventional hybridization technique maintained its tempo all along. However, there are certain disadvantages associated with this technique in that it is not applicable in many cases when the introduction of variability is from a species belonging to a different genus. Inter-specific introduction of variability is hampered in most cases by the development of pre and post-zygotic incompatibility barriers between the concerned species, which makes either the embryo totally incompatible with the endosperm resulting in abortion of the former or the fusion of gametes to form the zygote impossible.

It is also very well conceived that conventional form of crop improvement is time consuming. Starting from the generation/introduction of variability, it takes at least 8-10 years to come to the level of releasing a variety for commercial cultivation. In the present day scenario, such a long duration is unacceptable. Apart from this, a majority of the plant species do not allow themselves to be amenable to the introduction of genetic variability through conventional hybridization because of the fact that they are either vegetatively propagated or they take many years to come to the following stage (plantation crops). Hence, for generating variability through conventional hybridization in these crops one must resort to utilize alternative strategies.

The technique of mutagenesis relied on the use of mutagens that attack the nucleotide bases directly or indirectly causing changes in the target DNA. However, the use of this technology was somewhat indiscriminate. Hence, it became unsuccessful, with the advent of Plant Tissue Culture, another technique came into existence for generating variability i.e. somachional variation, the variation that has been observed in the regenerated plants and their progenies. The whole set up of tissue culture acts in a way similar to mutagenesis generating changes in gene sequences and chromosomes.

## In vitro crop improvement through genetic transformation

Introduction of foreign genes into plant cell genomes and genetically manipulating crop plants to suit human needs is one of the most fascinating perfected in a routine way, the introduction of alien genes into crops has to keep pace with it. However, gene transfer to some of the commercially important crop plants like legumes is still in infancy as the tissue culture/ transformation protocols are being developed slowly.

## Crop genetic engineering

Through the different approaches, major crop plants of importance viz., rice, maize, sugarcane, wheat, rapeseed, tomato, soybean, potato, ornamental plants were genetically engineered for value addition. The characters that have been engineered included resistance to herbicides, insects, male sterility, oil metabolism, fruit ripening, starch metabolism, nutraceuticals etc. Legumes were thought to be recalcitrant to in vitro regeneration and genetic transformation. Increasing emphasis has been placed on transforming these crops and legume crops like groundnut (Sharma and Anjaiah 2000) and chickpea (Krishnamurthi et at 2000) were transformed with high efficiency, recently.

## Genetic manipulation of crop plants

Genetic manipulation generally results in value addition in a crop plant for any genetic manipulation, appropriate crop plant for an appropriate trait should be chosen. Present state of the art technology of genetic engineering aids in the cloning of desirable genes from any organism and moving genes across kingdoms for producing desirable transgenic plants.

## Essentially, genetic manipulation involves the following steps

- Cloning of desirable genes for value addition and promoters (DNA fragments that drive the expression of genes hooked onto them) suitable for tissue specific expression.

  Cloning desirable promoters in genetic manipulation is of paramount importance in crop genetic manipulation. Use of constitutively expressing promoters for driving the expression of genes would render the transgenic plant commercially non-viable as the plant produces the protein encoded by the alien gene in tissues, where its production is not desirable. For example, if the choice trait is control of pollination in the form of male sterility, the alien gene should express only in the target tissue i.e. tapetum that nourishes the developing microspores. Expression of such a gene in a constitutive manner might be fatal to the plant in such a case. Hence, for controlling male sterility, the target gene should be hooked onto a promoter that is active in tapetum only. Like-wise, if the target trait is fatty acid metabolism, the genes should be expressed in developing seed and so on.

- Developing desirable vectors i.e. vehicles for moving expression cassettes comprising desirable genes and promoters across species.

- Proper marker genes that aid in the selection of transgenic plants. The marker gene expression in transgenic plants helps discriminate between the transformed and control plants. To illustrate, normal plants and plant cells cannot survive on an artificial medium containing the antibiotic kanamycin. If a gene called npt II (neomycin phosphotransferase) is expressed artificially in plant cells through genetic manipulation, transgenic plants and cells from them would tolerate the antibiotic kanamycin and grow comfortably on a medium containing the antibiotic. Hence, markers genes are essential in genetic manipulation.
- A method by which the desirable gene-cassettes are transferred to the target plant species.
- And finally a method that helps transformed plant cells regenerate whole plants by way of tissue culture.

  There are quite a few methods for introducing genes into plants. Of these, the vector based Agrobacterium medicated method of gene transfer and direct DNA delivery using a 'gene gun' are commonly employed.

The first genetic transformation system developed for plants was based as Agrobacterium tumefaciens based vectors. This bacterium was initially found to cause a disease in plants called 'crown gall' (Smith and Townsend 1907). Decades of research helped Scientists understand the mechanism of crown gall formation, which is a neoplastic diseases and apply the principle of nature genetic engineering successfully to transfer foreign genes to crop plants.

The soil bacterium A. tumefaciens has the exceptional ability to transfer a piece of DNA from the plasmid that it harbours called the tumor inducing plasmid (Ti-plasmid) into the nucleus of the recipient plant cells where it gets stably integrated (into the genome), transcribed and proteins made to its order resulting in the crown gall disease (Nester et al, 1984). The DNA fragment that gets transferred to the plant genome is called transferred DNA or (T-DNA). The native T-DNA contains genes responsible for crown gall formation (coding for enzymes that are involved in production of plant hormones) and genes responsible for the synthesis of opines i.e., condensation products of a sugar and an amino acid, which serve as a source of carbon for the bacterial cells. The mechanism by which the T-DNA is excised from the Ti-plasmid and transferred to the plant genome is well understood. The genes involved and the proteins coded by them are well characterized. These 'virulence' genes are borne on a 35 kb region of the Ti-plasmid called the 'vir' region. The proteins coded by them make a nick in the T-DNA that is placed between the 'border' sequences, the right (RB) and the left border (LB) (Zuopan and Zambrysky 1995, Gelvin 2000). Plant genetic engineering became a reality when the fact that the genes on the T-DNA do not play any role in the transfer of DNA from the Ti-plasmid to the plant genome and that any DNA fragment placed between the right and left border gets transferred to the plant cell (Zupan and Zambrysky, 1995, Gelvin 2000).

The above discussion brings out the following facts about the Agrobacterium mediated transformation.

- Tumor or crown gall formation is a process of transformation in which plant cells receive a transferred DNA fragment from the Ti-plasmid that resides in the bacterium.
- Genes present on T-DNA do not have any role in the transfer process; rather they get transcribed in the plant cell after getting integrated into the genome.
- Foreign DNA positioned between border sequences replacing the normal T-DNA also can get transferred to the plant cell.

The process of a genetic transformation based on Agrobacterium tumefaciens is relatively simple. Regenerable in vitro explants are simply treated with the bacterial cell suspension, blotted dry and are incubated on a semisolid nutrient medium of suitable composition. This process is called co-incubation. During this process, the different genes of the bacterial chromosome and 'vir' genes on the Ti-plasmid get activated, T-DNA excised from the Ti-plasmid and transferred to the plant cell. The whole process generally lasts 16-20 hours. Hence, co-incubation is generally allowed for 24-72 hours for different systems. The most important pre-requisite in Agrobacterium mediated transformation is the ability of the treated tissue to withstand co-cultivation with the bacterium and regenerate shoots. This method has been successful in transforming a very large number of dicot species and an increasing number of monocot species with a variety of alien genes for manipulating different morphological attributes.

To circumvent the problems in regenerating difficult crop species in culture and the problems of in vitro induced variations, efforts have started in the direction of avoiding tissue culture after co-cultivation with agrobacteria. These methods are in plant methods and use organized tissues. The first species to be transformed in plants was the model plant Arabidopsis thaliana (Feldman and Marks 1987) with the treatment of germinating seeds using the agrobacteria. After this effort, a variety of methods for infiltrating agrobacteria into the intercellular spaces of different tissues were successfully utilized (Katavic *et al* 1994, Clough and Bent 1998). Similar protocols of agrobacterial infiltration into intact plants at different stages were applied to Medicago truncatula (Trieu *et al.*, 2000) and non-tissue culture based agrobacterium co-cultivation of groundnut (Nalini and Sankara Rao 2000).

Besides agrobacterial co-cultivation, organized tissues were also used for producing transgenic plants using the method of electroporation. The method of electroporation was used earlier for producing transient pores in the protoplast membranes with the help of electrical pulses. The method of tissue electroporation (in contrast to protoplast electroporation) was successful in some species lupin (Chowrira *et al.*, 1995) and sugarcane (Arencibia *et al.*, 1995). At around the same time when Agrobaterium-plant interaction was being unraveled to pave the way for plant genetic engineering through different vectors, the other methods were also developed. These mainly included direct delivery of DNA to protoplasts through electroporation (Shillito *et al.* 1985, From *et al.* 1985, Lorz *et al.*, 1985), uptake of naked DNA mediated by polyethylene glycol (Uchimiya *et al.* 1984), micro projectile mediated delivery of DNA, popularly known as gene gun (Sanford 1988), floral tiller macroinjection of DNA, (De la Pena *et al.*, 1987) Protoplast based systems were initially the choice of Researchers in cereal transformation (for eg.Arencibia *et al.*, 1994). Since protoplast regeneration is technology intensive, protoplast mediated transformation had been replaced by Agrobacterium vectors and the gene gun method. All important crops viz., sugarcane (Arencibia *et al.* 1998), wheat (Cheng *et al.* 1997), rice (Hiei *et al.* 1994, Rashid et al. 1996), maize (Ishida *et al.* 1996) etc are now in a position to be transformed by the above methods.

## Transgenic Crop Plants

Transgenic crop plant production at commercial level has become a reality now. Latest estimates show that transgenic crops were grown on about 58.7 million hectares around the world. Of this area, USA (39.0 million ha), Argentina (13.5 million ha) and Canada (3.5 million ha) accounted for nearly 56 million hectares. Major crop plants soybean, maize, cotton and canola accounted for the entire area under transgenic crops and others like potato, squash and papaya have just entered the fray.

The characters, that received attention for genetic manipulation, are herbicide tolerance and insect resistance through the deployment of Bt genes (from *Bacillus thuringiensis*) that accounted for nearly the entire area of transgenic crop cultivation. Virus resistance in cucurbits and other characters accounted for the remaining area (of less than 0.1 million hectares). Apart from the above listed crops, significant work is already being carried out on commercially important crops like chicory, linseed, melon, rice, sugar beet, tomato, wheat, ornamental like carnation. In carnation, genes for delayed senescence (aminocyclopropane cyclase synthase – -ACC synthase) and sulfonylurea type herbicide tolerance (acetolactate synthase – als gene from tobacco) were introduced. The most common traits, herbicide tolerance and insect resistance were the favorites for the multinationals and they have paid a lot of attention in manipulating these traits. Since labour is very expensive in the western countries, it became imperative for the multinationals to look for genes that impart tolerance in crop plants to different herbicides. Each multinational concentrated on manufacturing a Trade Mark herbicide like 'Roundup', Basta', etc and started looking for the corresponding resistance gene. Hence major crop plants like soybean, maize, canola and cotton have transgenic with genes for herbicide tolerance incorporated in them. Similarly, Bt genes from different Bacillus thuringiensis strains were cloned and characterized by the different laboratories across the world. Apart from herbicide tolerance, canola oil has been modified using the genes for the enzymes that are involved in fatty acid metabolism.

Earlier, genetic manipulation in crop plants was successful when the target trait is controlled by one or two genes. However, now scientists are successful in introducing sets of genes controlling pathways also making complicated genetic manipulation a reality. A glaring example is 'Golden Rice' from Prof. Ingo Potrykus' lab in Switzerland, which has seven genes controlling vitamin A and iron accumulation in the seeds.

The first generation transgenic crops commercialized during the period 1996-2000 had herbicide tolerance and insect resistance as independent or stacked genes controlling weeds and target insects, which together can reduce crop yields by 24 per cent. These manipulations have produced significant environmental benefits. In fact, Bt maize is less susceptible to infection by fusarium, which produces fuminosin, a carcinogen. The development of these transgenic has resulted in the reduced application of herbicides and insecticides and offer efficient weed and insect control. There is an increasing body of evidence that transgenic crops are delivering significant economic benefits (ISAAA Brief 2003). Apart from herbicide tolerance and insect resistance, the other traits targeted are disease resistance, pollination control mechanisms, metabolic engineering, value additions, traits important for ornamental crops etc.

## Terminator Technology – Control of Seed Germination

This technology changes the genetic architecture of the plant cells in such a way that the seed produced on the transgenic plants regenerated from genetically modified cells and distributed to a farmer would germinate, grow normally into plants and set normal seed. However, the seed set on the plants grown by the farmer would not germinate. This is because of a cascade of events that would be occurring in plants growing from the seed supplied by the company to the farmer and a chemical treatment given to the seed at the time of supply to the farmer.

This technology is based on a set of genes incorporated into the transgenic plant

- A gene coding for a toxic substance lethal to the developing embryo. This gene is linked to a blocking sequence whose presence between the promoter and the lethal gene inactivates the lethal gene from expression.
- A recombinant gene, whose expression results in the removal of the blocking sequence that is preset between the promoter, that is active at late embryogenesis stage, and the lethal gene. This removal of the blocking sequence leads to the production of the lethal substance during embryo development and lethality.
- A repressor gene that codes for a protein suppressing the lox gene. However, this repressor gene is under the control of a repressible promoter. A chemical trigger tetracycline, an antibiotic controls the promoter.

Normally, the repressor gene is on and the recombinate gene is switched off in the first generation transgenic plants developed by the Company and they produce normal seed that germinates well. In this case, the toxic gene is switched off embryogenesis specific promoter and the lethal gene. This seed will be supplied to the farmer after applying the chemical trigger, the antibiotic tetracycline. This seed will germinate normally on the farmer's field. However, treatment with tetracycline, results in the repression of the promoter of the repressor gene resulting in the gene turn off. The absence of repressor protein in turn switches the lox gene on. The lox gene product then removes the blocking sequence between the promoter and the lethal protein gene thereby switching on the lethal gene. However, the promoter is active only in the late embryogenesis stage and the developing embryos become non-viable. Hence, next generation seed will not germinate. This technology does not offer any benefits to the farmers nor to the consumers. The only advantage would be with the commercial seed companies forcing farmers back to the market for buying the seed repeatedly.

## Ethical Issues

Bioethical issues do not matter much in crop biotechnology as compared to medical biotechnology. Bioethics would come into picture only when DNA fragments are taken from obnoxious organisms and inserted into plants for expression. The question would be whether public would accept such food/plant products from genetically modified organisms. Major problem in crop genetic engineering is the horizontal transfer of alien genes back into wild germplasm, which would result in the development of super-weeds and microbes becoming resistant to antibiotics. Hence, proper precautions are to be taken into consideration with regard to bio-safety.

## Future Prospects

Future looks to be rosy for Crop Biotechnology. More and more commercially important plant species will be genetically modified for value addition. There will be advancements in the field of plant bodies and plant vaccines and we hope to see these plant based products in the next three to five years. Nutraceuticals is fast becoming a reality with medically important plant products going to hit the market. Already, some products are in advanced clinical trials and would be available for human consumption soon.

India also seems destined to enter into the area of commercial production of transgenic crops. The Government of India has approved the BT-Cotton for commercial cultivation.

There is indeed an urgent need to educate the people about the myths they have about Biotechnology. This is particularly important when the global population is on an increasing trend and cultivable land remains static or getting dwindled. Moreover, yields have reached a plateau in a majority of our important crops and the Scientists are left with no option other than resorting to biotechnology. For India, which is a late entrant on the scene of Crop Biotechnology, the burden of proving the safety of the GM foods has already been by the advanced countries in many cases. Lot of experimentation on the safety has already occurred and the products were certified in many cases. Over and above this, the environment in which the man is living also gets cleaned up because of the reduced use of pesticides, herbicides etc.

### *Reference*

Biotechnology and Development Monitor Issue 35,1998

Biotechnology and Development Monitor Issue 32,1887.

ISAAA Brief No.27, 2002.

# UNIT - IV

## Partnership and Communication Strategies

# 19

# Collaboration Between Public and Private Sector in Biotechnology R & D

**N.P. Sarma[1] and K.B.R.S. Visarada[2]**

A big challenge to agriculture in the present quarter century is to balance global food and nutritional security with environmental safety. Demographic studies forecast an increase in world's population from current 6 billion to 8.5 billion by the year 2025. It is obvious that the progress in productivity increase and agricultural production witnessed in the last few decades through efforts in plant breeding and agronomic practices like fertilizers, irrigation and crop protection will not be sufficient to meet the growing food demand. Furthermore, agriculture is the engine of growth in a developing country's economy as it provides not only food, but also employment, income and livelihood security to the majority of the

[1]Former Principal Scientist, Directorate of Rice Research, Hyderabad - 500 030.

[2]Scientist, National Research Center for Sorghum, Hyderabad - 500 030.

population. Thus, investments in agricultural research and development (R&D) are surest ways to improve the economic well-being of rural population. To sustain the pace of agricultural growth there is a need for new technologies to increase productivity. Crop Biotechnology has a role to contribute effectively to the cause. Biotechnology, however, should not be viewed as a substitute for traditional methods of crop improvement. Integrating biotechnology tools into traditional breeding programs could substantially enhance the efficiency of agricultural R&D. Through molecular breeding targeted transfer of desired genes into crops or introduction of novel traits from biological gene pool are feasible. Agronomic or input traits such as tolerance/resistance mechanisms against abiotic and biotic stress could enhance crop output in diverse agro-ecologies including the marginal lands. Quality or output traits including enhanced micronutrients in grain could improve nutrition and health of consumers. There is a tremendous potential yet through exploitation of advances in genomics. Despite the great potential, Biotechnology developments are so far mostly focused on large scale commercial agriculture in developed countries. About 80 per cent of all research investments in agricultural biotechnology are made by the private sector whose interest centers around important global crops and traits rather than orphan crops in less developed countries. Thus, technologies as well as important tools are becoming increasingly proprietary with relevant IPRs held by the corporate sector. International investments in agriculture by the public sector are dwindling. Of late, the public sector funding is shrinking more and more. At the same time introduction of patent rights is paving path to more privatization of technologies. Thus the public sector industry is threatened by the dynamic and expanding private sector. Biotechnology revolution will bypass the poor unless suitable policy mechanism is developed and implemented to improve developing countries access to the required innovations.

Developed countries have been benefited from agriculture research and development both in public and private sectors but developing countries by and large depend on public sector support from National Agricultural

Research Systems (NARS) and from International organizations such as the International Agricultural Research Centres of the Consultative Group on International Agricultural Research (CGIAR). Research domains that are of less interest to private sector have traditionally been addressed by public sector institutions. Biotechnology R&D is a resource intensive endeavor both in terms of human capital as well as financial resources. In the face of declining investment for agricultural research by the public sector organizations, developments in technologies generally and biotechnology in particular is increasingly spear-headed by private sector endowed with modern infrastructure and critical mass. In addition to resource crunch, there are also other shortcomings limiting the public sector R&D in Biotechnology. The IPR issues and the patents held by the corporate sector make it almost impossible to access even the basic tools without interacting with the private sector organization. Creating innovative partnerships between public and private sector based on comparative advantage is the way forward to ensure the development of technologies that can address the needs of poor farmers in developing countries.

## Public Sector Research in Biotechnology

### (a) International Agricultural Research Centers (IARCs)

The public sector and other International and National charitable foundations supported the crop research that led to phenomenal increase in yields in Asia and Latin America through evolution of dwarf wheat and rice varieties. Substantial public sector agricultural research still has a major role to play despite declining funding support. The CGIAR system comprising of 16 International Agricultural Research Centers with interest in specific crops — wheat and maize (CIMMYT, Mexico), rice (IRRI, Philippines), potatoes (CIP, Peru) millet, sorghum, chickpea, pigeon pea and ground nut (ICRISAT, India), substantially contribute to basic research and its strategic application.

The IARCs can play an important role in adopting the tools of modern biotechnology to the needs of Agriculture in developing countries. Most IARCs are involved in Biotechnology to varying degree. The CGIAR centers have progressively invested in biotechnology and applied the tools of Biotechnology to furthering CGIAR mission–alleviation of poverty and protection of natural resources for sustainable food security. In the advent of new advancement in molecular biology and genomics, a CGIAR panel considered the future strategies and related activities in biotechnology. In order to be a serious player in biotechnology, an international center needs to have:

- A critical mass of scientific expertise
- Skills to acquire new technologies from industrialized countries under suitable licensing agreements
- Capacity to integrate new technologies into the ongoing R & D programmes

## (b) Food and Agriculture Organization (FAO)

FAO is promoting the application of agricultural biotechnology among developing countries and currently assisting 30 developing countries through implementation of specific projects through research and information networks, such as UNDP/FAO/UNIDO Regional sub-program on Biotechnology and Biodiversity of the Farmer centered Agricultural Resources Management (FARM). Biotechnology networks have been established to disseminate information and training. Among these are Plant Biotechnology network (REDBIO) and Cassava Biotechnology network. Plant Breeding and Genetics section of joint FAO/IAEA Division of Nuclear Techniques in Food and Agriculture supports the use of biotechnology for plant improvement through technical cooperation projects in tissue culture and coordinated research programs.

## (c) USAID and International Agricultural Biotechnology Program

US Agency for International Development (USAID) while designing new programs in agri-biotechnology considered a number of factors about involvement of private sector in collaboration with US, international and developing country research institutions. This collaboration sought with private sector is viewed as means to access both research tools developed by the private sector and specific technical expertise. USAID designed a program integrating all aspects associated with the dissemination and application of biotechnology, particularly management and technology transfer issues with biotechnology research and training. USAID has supported several public-private sector collaborative research programs largely through the Agricultural Biotechnology Support Program (ABSP) led by Michigan State University. ABSP represented partnership between a number of US Universities and a number of developing country companies, the IARCs and National Agricultural Research System (NARS). Some examples of public-private sector partnerships supported by USAID include:

- Monsanto and Kenyan Agriculture Research Institute (KARI)
- DNA plant technologies and Costa Rican and Indonesian owned tissue culture companies
- ICI seeds and Central Research Institute for Food Crops (CRIFC, Indonesia)
- Pioneer Hi-bred and Egyptian Agricultural Genetic Engineering Research Institute (AGERI)

At the national level important factors that contribute to decline in public sector research is brain drain and less commercial approach. NARS have contributed greatly to the green revolution and in attaining self sufficiency in food. An additional contribution can come only through market and export oriented outlook. Lack of avenues for entry of young talent and resigning the public sector positions, for more remunerative offer in the private sector or abroad are on the increase at the national scenario. A recent trend in developing countries is that private firms tend to provide more investment and more support per scientist in research than public sector.

## Private Sector in Biotechnology R& D

Most of the agricultural biotechnology products to date have been developed by private sector. The first wave of agricultural biotechnology–insect and virus resistance, herbicide tolerance and post harvest handling has been commercialized in crops like soybean, maize, cotton, canola, papaya and several other vegetables. In the industrialized countries the new biotech/ life science companies (the big trans-nationals) through a process of acquisitions and merger are able to combine crop biotechnology with agrochemical and seed production. AgrEvo, Dow, Dupont, Monsanto, Novartis and Syngenta have dominated the application of biotechnology in agriculture. So far, these companies focused on developed country markets where potential sales are large, patent protection assured and risks are lower. Some of them even conceptualized a technology (GURT) that would ensure seeds harvested from GM crops are sterile and cannot be used for raising next season crop, to keep the recurrent demand for their product. There is less interest in small farmers and food crops because the returns are low as 80 per cent of farmers in the developing countries keep their seed for next years crop. Why the private sector is only interested in a few crops and traits is obvious because of enormous investment for developing transgenics and other biotech products. This problem of high investment is compounded by the long period of product development that requires 6-10 years.

By 2003 transgenic crops of canola, cotton, maize, soybean etc., have been widely adopted by several countries. Development of these transgenic crops has taken many years involving technological breakthroughs. The early phase till first commercialization almost took two decades. No doubt, most of the transgenic varieties grown now, are the products of private sector effort. In the product development-evaluation phase that involves large scale field testing, diversification of varietal backgrounds for wider adoption and regulatory processes and registration for official approval and frontline demonstrations prior to release, the private sector needs active collaboration with NARS (extension services of SAUs, Seed Certification agencies etc).

## Role of public and private sector in agricultural biotechnology for developing countries

At present public sector research is dominating in most developing countries. But with the introduction IPR regimes there is a gradual shift towards privatization and there is always an element of doubt on how and how much contribution of each sector should be allowed. In India contribution of agriculture is significant in the economy of the country and farmers here are subsistence oriented (Morris & Ekasingh, 2002). Till now needs of the agricultural sector have largely been met through intense efforts of National Agricultural Systems.

The system of patenting - genes, expressed sequence tags (ESTs), gene transfer techniques, genetically improved lines (GI) is justified in terms of protecting IP and stimulating innovation. For developing countries, however it is denial of the innovations, for the benefit of poor. A solution to this is in adopting a mixed approach through public–private sector partnership by providing economic incentives for private sector research and at the same time ensuring public sector access to new technologies. Several initiatives are already underway to build partnerships between public and private sectors. Some of public–private initiatives are: International Service for Acquisition of Agri-biotech Applications (ISAAA), establishment of public-private sector consortium by CAB International to support Global electronic crop protection compendium (CPC). ISAAA's mission is to transfer and deliver appropriate biotechnologies and products from the private sector to developing countries.

Research programs of not commercially exploitable value do not attract private sector investment and if these are for public good they need to be addressed by NARS and IARCs or other agencies–non-profit foundations like M.S.Swaminathan Research Foundation and the AP-Netherlands Biotechnology programme etc. For example, for decades almost all rice research was carried out by the national public sector centers. The Rockefeller Foundation sponsored an International Program Rice Biotechnology (IPRB) as early as 1984. An investment of 100 million US $ over a period of 15 years has made rice as model for monocots and top among the cereals. The private sector has played little role in development and delivery of rice technology in the tropics till the advent of hybrid rice development and attraction of biotech opportunities.

According to Morris & Ekasingh (2002) the public plant breeding programs should reinvent themselves to perform five essential functions that private sector is likely to perform at socially sub-optimal levels

(i) Genetic improvement for specific products, crops and traits

(ii) Conservation of genetic resources

(iii) Training of plant breeders

(iv) Varietal testing and evaluation

(v) Biosafety regulation

Genetic improvement efforts should be focused to target selected products, crops and traits. Public sector breeding programs should take up work in existing cultivars and modify them to generate broad social benefits e.g., Golden Rice, QPM (Quality Protein Maize). Private sectors can have little commercial returns for enhancing the nutrient levels, and subsistence farmers may not be in a position to pay for royalties/costs.

Many NARS centers and IRRI are the major sources of genetic diversity and basic biological knowledge of rice. Biodiversity provides the feedstock for Biotechnology industry. A good example of value of gene discovery is the *Xa21* gene that enhances resistance to Bacterial leaf Blight. These are freely shared for further innovation and cloning. The genetically improved lines near isogenic lines and gene pyramids are made available at zero royalty. Breeding for host-plant resistance can be a high priority in public sector thereby reducing the need for pesticide use for environmental benefit. Similarly development of parental lines can be taken up at the public sector due to their accessibility to wide germplasm. Improvement of local varieties for specific traits can be addressed through public sector research. Another case in point for forging public and private sector collaboration is rice genomics. In order to foster advances in frontier areas of science the expectation of both the partners of collaboration need to be recognized. The challenge is to develop a shared vision in research that will provide the public sector access and freedom to use tools of biotechnology and at the same time offer sufficient incentives for private sector to innovate, develop and deliver new technologies. Being a crop of poor Asian farmers rice was not as attractive as Maize for the corporate sector to invest in research in the past. With the spurt in rice research through the applications of molecular tools and demonstrated commercial feasibility of hybrid technology, rice has attracted corporate investments. Expectedly, such investments aim at reaping the benefits through products for profit and denial to poor farmers. Such a collaboration is envisioned in rice functional genomics. The need for diverse genetic resources (held by NARS and IARCs) and expertise in functional genomics requires a common platform of partnership where information and genetic sources can be broadly shared. The announcement of the release of rice sequence data and donation of proprietary technology and the product for the development of beta carotene rich rice are the signs of private sectors willingness to share their products of investment and make them available to developing countries. The pattern sharing rights in the public-private sector partnership is that genetic resources for functional genomics will be made available for both the parties under a Material Transfer Agreement (MTA). MTA facilitates recipients to obtain patents on genes discovered through the use of material but obliged to make available rights under those patents at a reasonable royalty for commercialization and zero royalty for subsistence farming. The high investment required for biotechnology research is the cause for high cost price of the end product such as GM seed. Economic returns through cost price cannot be expected if the end user is the less privileged sector of people in the societies. Mandate of the public sector research should focus on the areas that are likely to be neglected by private sector due to lack of commercial incentives. An example of this is the "Golden rice" that synthesizes Vitamin A in the endosperm. End users are malnourished people of the world who are rice eaters. The Golden rice thus produced is intended to be released in public domain for human welfare. On the other hand, there are funding agency obligations that affect the freedom to separate. Low incentive crops which are primarily cultivated with low inputs by resource poor farmers can be taken up by public sector systems. Decision of the Andhra Pradesh Netherlands Biotechnology program to make available their technologies freely to the end users, is an important step to show how the public sector research supported by philanthropic organizations can be used for public good. However, declining talent and expertise in the public sector organization, low incentives for challenging jobs, less freedom of operation, declining financial support are some of the factors that need to be weighed carefully, before venturing on advanced technologies in a public sector. Three examples of public and private sector collaboration in rice are elaborated here:

### (i) Saga of *Xa21* gene

An example that embodies application of rice biotechnology tools was the discovery of the bacterial blight resistance gene *Xa21*. It contributed to the understanding of the basis of plant disease resistance and the genetic mechanisms. Initially, rice lines of *Oryza longistaminata* originating from Mali and Africa were identified to carry broad spectrum resistance to bacterial blight. Through the interspecific hybridization the gene *Xa21* was transferred to cultivated rice. Further X*a21*gene was mapped, sequenced and engineered into a susceptible rice genotype. With the application of marker assisted selection using RFLP and PCR *Xa21*was pyramided with other *Xa R* genes. *Xa21* gene was genetically transformed into cultivated rice variety IR72. Field trials were conducted in China, India, Indonesia and Philippines for pyramided *Xa* genes, including *Xa21*. A restorer line developed with *Xa 21* gene is used in hybrid rice. Thus the saga of bacterial blight disease resistance gene, *Xa21,* epitomises the international collaborative research, training and capacity building across the globe and illustrates the successful case of extension of knowledge generated to production of improved rice varieties. A patent was obtained for the gene in 1999. Genetic Resources Recognition Fund was founded and it was decided that the licence fees and royalities will be used to assist science capacity building in developing countries.

### (ii) Development of Golden rice

Private sector contributed to the bulk of research and development in wheat and maize, whereas public sector has accounted for most of the technological advances in rice production. The development of Golden rice was made possible by the fund support received by two agencies, the Rockfeller foundation and the European community.

The Rockefeller Foundations's support was free of obligations, while European community has an industrial partner, Zeneca. Current legal status of Golden Rice Project has a two track obligation (i) non commercial (humanitarian) and (ii) commercial (Fig. 19.1).

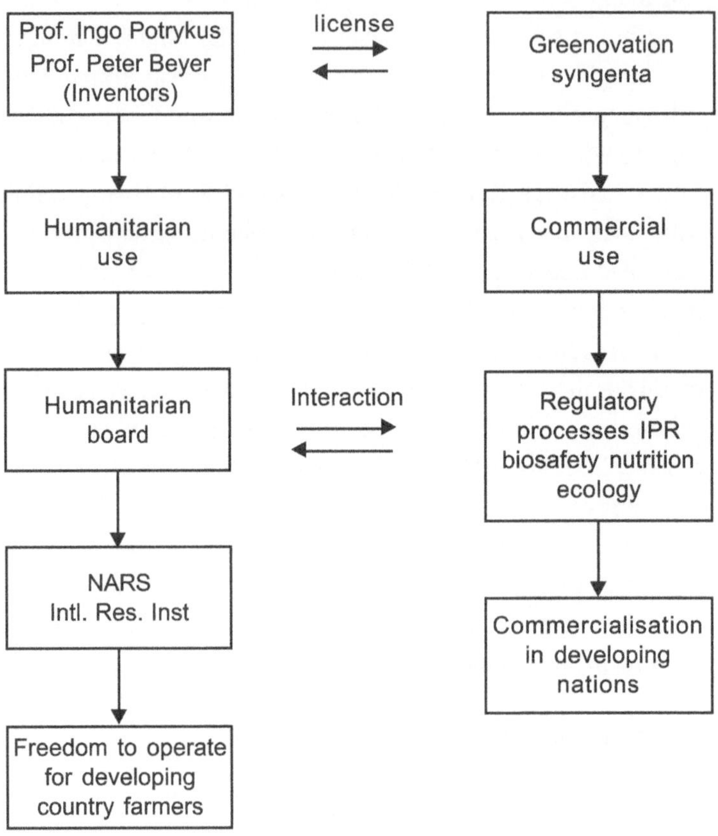


**Fig. 19.1** The legal status of the Golden Rice Project (Adopted from Beyer *et al* ., 2002)

Nearly 44 patents are involved for use of Golden Rice and these are mostly held by the private sector and a few by Universities. An agreement was reached between the inventors (Professors Ingo Potrykus and Peter Beyer) and Syngenta to serve mutual interests of the technology–for humanitarian use and commercial application. An advisory panel, called the Humanitarian board has been set up for the purpose. Non-commercial license for use of the technology can be made available to National and International Research Institutes through the Humanitarian Board. This will facilitate the transfer of beta carotene trait into local varieties by breeding or transformation by countries like India, China, Indonesia, Vietnam and Philippines, who seek licenses from Humanitarian Board.

### (iii) International Rice Genome Sequencing Project (IRGSP)

Private sector participation in the International Rice Genome Sequencing Project is another example where company resources (draft sequences of the rice genome) were pooled with the sequences produced with public funds, resulting in quicker release of rice genome. The International Rice Genome Sequencing Project (IRGSP), a consortium of publicly funded laboratories, was formed in 1997 to obtain a high quality, map-based sequence of the rice genome. It is currently comprised of ten members: Japan, the United States of America, China, Taiwan, Korea, India, Thailand, France, Brazil, and the United Kingdom. Monsanto and Syngenta, two giant private sector corporations, announced to make available the genome sequence for IRGSP and others who need it.

Another example of public-private partnership is the Human Genome SNP (Single Nucleotide Polymorphism) Consortium that pooled the public and private resources to complete the human SNP database. Celera Genomics established a reference database on human SNPs, which includes those screened from public sector databases. Mouse Sequencing Consortium (MSC) consists of six public sector institutes and three private companies. Information generated through these projects is crucial for basic biomedical research.

Summarizing the role of public and private sector involvement in developing GM technologies, the Royal Society of London along with other science academies of the world recommended that (i) governments should fully recognize that there will always be public interest research requiring public investment even in the market–driven economy–it is imperative that public funding of research in this area is maintained at least at its present level in both the CGIAR and National Research Institutions; (ii) governments, international organizations and aid agencies should acknowledge that plant genomics research is a legitimate and important object for public funding, and that the results of such research should be placed in the public domain; (iii) innovative and vigorous forms of public private collaboration are urgently required if the benefits of GM technologies are to be brought to all the world's people (iv) incentives are needed to encourage commercial research companies to share with the public sector more of their capacity for innovation and (v) care should be taken so that research is not inhibited by over-protective intellectual property regimes.

## Bibliography

Toole, JCO, Toenniessen, GH, Murashige T, Harris RR and Herdt RW The Rockefeller Foundation's International Program on Rice Biotechnology

Morris M and Ekasingh B 2002–Plant breeding research in developing countries: what roles for the public and private sectors? In Agricultural Research Policy in and Era of Privatization. Eds Byerlee D and Echeverria RG. 199-225.

Beyer P al-Bablii , Ye X, Lucca P, Schaub P , Welsch R and Potrykus I 2002-Golden rice: introducing the ß-Carotene biosynthesis pathway into rice endosperm by genetic engineering to defeat vitamin A deficiency. In Experimental Biology Symposium, Plant Breeding: A new tool for fighting micronutrient malnutrition. April1,2001, Orlando, Florida, US.

# 20

# Communication Strategies and Linkages for Agricultural Biotechnology Sustainability

*M.N. Reddy

Biotechnologies are likely to suffer the fate of non-adoption by the farmers and users. Even the technologies that have been hailed as solutions and innovations have suffered the non-adoption due to inadequate public awareness, inappropriateness to the farming system or agro-ecology and high cost/ low profitability. People have prejudices that biotechnologies are not suitable to stressful or unpredictable conditions. The R & D focus is on rapid pace of research and technology development with least importance to developing of common people's knowledge. Hence, appropriate communication strategies have to be designed to increase public awareness.

* Director (Agril. Exten.) and Constultant (NATP), National Institute of Agricultural Extension Management (MANAGE), Rajendranagar, Hyderabad - 30.

Due to diversity among Indian people, particularly geographic expanse, economic, cultural and social difference, strategies for communication of Biotechnology cannot be uniform or one-shot-go. The challenge includes differences in attitudes, varied levels of education and understanding of the issues, prejudices and preconceived notions, lack of coordination among groups offering services, lack of resources for communication and strong opposition to Biotechnology. The gap between rapid pace of scientific research and development and slower pace of societal learning needs to be bridged.

## Frequently raised concerns about Biotechnology (Traynor, 2001)

- Fear of the unknown due to little or no knowledge or no understanding of technical details and low levels of understanding about science in general amplified by recent cases of unanticipated harmful consequences arising from other new technologies. For example, introduction of exotic seeds, use of DDT pesticides.
- Mistrust of government officials and multinational companies due to past experiences.
- Resistance to change shown by placing a high value on traditional farming methods; assertions that Biotechnology is not needed because other methods would work as well.
- Sense of unfairness, those who reap the benefits are not the ones who bear the risk.
- Inaccurate perceptions of the risk shown by a even demand for zero risk and proof of safety; reluctance to make trade-offs between risks, costs and benefits.

## Need for public awareness on biotechnology

- To provide food for one billion population, it will be necessary to expand food production faster than the rate of population growth. As further expansion of the area cultivated is impossible in most of the world's developing countries, the future gains in food production have to come from increased yield (Bunders, *et al* – 1996)
- Bioengineering techniques are powerful new tools to supplement pathology, agronomy, plant breeding, plant physiology and other current approaches. If crop Bioengineering techniques developed and applied in a manner consistent with ecologically sound agriculture, they could decrease reliance on a broad spectrum of insecticides which cause serious health and environmental problems.
- Biotechnological techniques could also assist in the development of crop varieties that are resistant to currently uncontrollable plant diseases.
- It has been proved that biotechnology revolution in agriculture can dramatically reduce the costs and increase outputs. It is based on the frontier areas of science and thus economically useful and viable. Biotechnological research cost is comparatively less in India than in the Western counties.
- The Biorevolutions as against the green revolution can help in saving resources spent on pesticides, fertilizers, irrigation etc., leading to low cost of production.
- Biotechnology has been identified as one of the key carriers of the next wave of technological development as that of Information Technology.

A comparison between Agricultural Extension and Biotechnology Extension is given in Table 20.1

**Table 20.1 Comparison between the Agricultural Extension and Biotechnology Extension**

| Keywords | Agricultural | Biotechnology (*) |
|---|---|---|
| Focus | Production | Market |
| Target Group | Farmer | Consumer, Farmer |
| Origin of technology | Field | Mostly lab |
| Time dimension | Time-tested | Recent Origin |
| Human Resources | *Labour intensive, *Para-extension workers | *Capital intensive, *Scientifically trained personnel |
| Linkages | Research-Extension-Farmer | Research-Extension-Farmer-Marketing-Consumer |
| Main stakeholders | Extension agency, Farmers, Policy Makers | Scientists, Industry, Traders, Financiers, Farmers Extension Consumers, NGOs, Policy makers, Representatives of farmers and women's groups. |
| Orientation | Service | Profit-making |
| Extension system | *Public Extension system, *Free | *Private Extension System, *Priced |
| Accountability | Service | Product |
| Credibility | High | Suspected |
| Understandability and adaptability | High | Low |
| Awareness among farmers and consumers | High | Low |

(* Mostly with higher form of biotechnology (GMOs etc)

**Table 20.2 Present status of Knowledge (K) Skills (S) and Attitude (A) amongst main stakeholders about Biotechnology**

| Stakeholders | Agriculture | | | Biotechnology | | |
|---|---|---|---|---|---|---|
| | **K** | **S** | **A** | **K** | **S** | **A** |
| 1. Research | H | H | H | H | M | M |
| 2. Policy makers | H | NA | H | L | NA | H |
| 3. Industry | H | H | H | H | H | H |
| 4. Public Extension | H | M | M | M | L | M |
| 5. Farmers | M | M | H | L | L | L |
| 6. Consumers | H | NA | H | L | NA | L |
| 7. Media | H | M | H | L | L | H |
| 8. NGOs | M | M | M | M | L | L |

(**H**-High; **M**-Medium; **L**- Low; **NA** - Not applicable)

## Consumer acceptance of any Biotechnology application will occur more readily when

- Most sought after technology applications (e.g., golden rice)
- Obvious benefits to the consumer (e.g., lower price, better quality produce etc.)
- Communications about the advancement are open, honest and free of Jargon
- The risks are virtually nil and research is available to support this claim; and
- The genetic modification is plant rather than animal based.

For planning and successfully carrying out the public campaign regarding the biotechnology applications, necessarily the pulse of the public should be studied from time to time. Research should focus on gathering reliable and timely indicators of the changing attitudes and opinions for designing the communication strategy. Following are some of the communication strategies developed for building awareness among important stakeholders. The strategies presented here are not comprehensive in nature, but may provide general guidelines and are based on relevant experiences in the field of Biotechnology.

## Communication Strategies for Research / Scientists

- Encouragement of Participatory Technology Development (PTD) rather than scientist centered technology so that the technology will be made easy for propagation and multiplication.
- Conduct of on-farm and adaptive trials by involving all the stakeholders
- Conduct of participatory forums to understand the technological needs of stakeholders and direction of research.
- Organization of conferences and symposia to exchange and share the development.
- Video conferencing for clearing the doubts with other scientists.
- Publication of scientific digests to disseminate the information at national /state level.
- Creation of web pages and network of scientists and organization contributing to the development of Biotechnology with e-mail Ids.
- Broadcasting of talks on radio / T.V in local language.
- Conduct of annual / biennial scientific forums at national / state level and provide incentives.
- Production of video films and CDs on advancements in Biotechnology.

## Communication Strategies for Policy Makers

- Constitution of multi-disciplinary teams of policy makers consisting of experts from the disciplines of Biotechnology, economics, environment, agriculture, sociology.
- Participation in stakeholders interface forum.
- Conduct of surveys to understand the pulse of all stakeholders, which would help in appropriate policy formulation.
- Constant interactions with the internet and surfing through the relevant web pages.
- Pilot test of policy prior to application on a state / national level.
- Production of relevant video films and CDs for updating the information
- Print media like newspapers, science journals etc.
- Organization of conference and symposia to exchange and share the development.
- Video conferencing for clearing the doubts with other scientists.

## Communication Strategies for Industry

- Constant linkage with scientific community to get updated on the latest technological developments.
- Scientific digest to understand the technological innovations.
- Attending conferences and symposia of scientists.
- Creation of web pages and network of scientists and organizations contributing to the development of Biotechnology with e-mail Ids.
- Conduct of surveys to understand the attitude of people.

## Communication Strategies for Public Extension

- Participation in Participatory Technology Development (PTD) process.
- Interface with scientists to get updated on the latest technical developments and give feedback to the scientists.
- Attending the scientific get-together.
- Actively participating in on-farm and adaptive trials.
- Attending trainings regularly
- Exposure visits within and outside states
- Publications like scientific journals and magazines.
- Production of video films and CDs on advancements in Biotechnology.
- Attending kisan melas that focus on Biotechnology.
- Creation of web pages and network of scientists and organizations contributing to the development of Biotechnology with e-mail Ids.

## Communication Strategies for Farmers

- Involvement of farmers in Participatory Technology Development (PTD) right from the identification of the researchable issues through various stages of final release.
- Demonstration through field trials.
- Organization of regular interface of all stakeholders involving farmers.
- Exposure visits to research centers and successful farmers' fields in districts where Biotechnology programmes are in progress.
- Development of resource persons among farming community to act as effective linkage between the stakeholders [e.g., through Farmer Organisations (FOs) Farmer Interest Groups (FIGs)].

- Usage of television, radio and newspaper through short films and slogans in local language on a continuous basis.
- Conduct of kisan melas / exhibitions by universities and research institutes to share the research outcomes and familiarizing the farmers with technologies.
- Poster displays with curious queries about Biotechnology in prominent places of villages / mandals (taluks) / districts headquarters.
- Orientation of Rural Agricultural Work Experience (RAWE) programme students of SAUs to spread the message of Biotechnology during their stay in village.
- Conduct of series of gram sabhas with a Biotechnology expert.
- Advertisement through displays during Jatharas, shandis etc., in rural area.

## Communication Strategies for Consumers

- Participation in Participatory Technology Development (PTD)
- Organization of Bio-forums for consumers with interface of research, farmers etc.
- Preparation of display windows at shops with Biotechnology products.
- Regular broadcast of messages on AIR and Television including the private channels.
- Regular advertisements in the newspapers with brief slogans.
- Regular advertisement on the back covers of general magazines.
- Display of attractive posters at the retail outlets and during jataras, shandies etc.
- Presentations in gram sabhas with Biotechnology experts to attract the rural consumers.

## Communication Strategies for Media

- Increase the Biotechnology community's awareness of the importance of cooperating with the media.
- Scientists and Biotechnology institutions should develop an active approach to the media to create a variety of background materials, briefing documents useful to various media.
- Exploration of opportunities to present information in a variety of local traditional media such as folk media, burrakathas etc.
- Encouragement of institutions to explore internships and other cross fertilization programmes between educational institutions and the media.
- Need for creation of media cell at state and national level for propagation of uniform messages through radio, television etc.
- Encouragement of professional scientific organizations to discuss communication issues and involve the media where appropriate.

## Communication Strategies for NGOs

- Participation in Participatory Technology Development (PTD) process
- Continuous interface with scientists
- Attending scientific get-togethers
- Promotion of Farmers Organization / Farmer Interest Groups (FIGs)
- Involvement in conduct of on-farm trials / adaptive trials
- Organization of exposure visits
- Publication of literature in local language
- Orientation to consumers
- Coordination with local media

Effective communication strategy on Biotechnology should create public and political confidence, encourage future investment, reduce fears and skepticism and improve the usage of Biotechnology products. The Biotechnology community should ensure access to the balanced and credible information so that people reap the benefits of Biotechnology by deciding that Biotechnology products would contribute to their well-being.

## LINKAGES

Biotechnology is using information about biological systems to develop techniques for commercial production of food, pharmaceuticals, environmental protection, biochemicals and clinical assays. It is expected to produce significant economic and environmental benefits as the result of the development of new varieties of plants & animals and enhanced agricultural productivity. In this process, there are many stakeholders in the chain and very important among them are consumer, researcher, extensionist, farmer and market mechanisms. The ultimate end user is the consumer, hence assumes higher proportion of importance. Consumer response and acceptance of Biotechnology tends to be application-specific and acceptance is more likely to be achieved if specific benefits can be realized by the consumer like lower price, better quality produce, with distinct and valued characteristics. In order to win over the consumers, the linkage chain needs to be thoroughly organized and address the issues concerned of each stakeholder.

### Consumer's opinions take shape, based on the following factors

- The technology's ethical ramifications
- Its impacts on the environment
- Potential health risks

- Its effect on the agriculture
- How it may affect future generations
- The extent to which the technology and its management by Government and industry are trusted by concerned consumers.

According to a Canadian survey, Apparent Consumer Preference for Biotech Applications (highest to lower acceptance) are :

- Application, which meet clearly identified social and personal needs
- Applications, which meet a high social need but where consumer benefits are less evident.
- Applications, which meet a social need but offer no value to the consumer.
- Applications, which offer some benefit to the consumer but have no apparent social value.
- Applications, which are perceived to offer no benefits to society or consumer and where the benefits apparently go only to the producer, with the consumer carrying most of the risk.
- Application, which involve transferring genes between animals and plants
- Any application involving the transfer of human genes.

The listing underlines that consumers in surveys placed more weight on benefits to society than to themselves as consumers, and that socio-ethical concerns weighed heavily in their responses. This suggests that the applications in the lower categories would have to generate significant benefits to society to achieve market acceptance.

## Influence of consumers on Biotechnology

- They attempt to influence the industry through market mechanisms by bringing their concerns about a product into their purchasing decisions.
- They participate in the process of providing public inputs for policy making to the government, as policies should reflect public concerns.
- They are the source of advice regarding consumer protection since they will be affected directly by many aspects or biotechnology.

By representing their own interests through consumer groups, they can provide input to industry in policy making.

Consumer role is consumption oriented, focusing on obtaining and evaluating information about different products and using this information to select and consume products, which will give greater benefits to them. Such choices may be based on factors related to the production process and product attributes and may be motivated by broad societal concerns as well as concerns of health, safety and consumption enjoyment.

## Utility of Linkages

Linkage mechanisms are used to channel information between groups and to conduct required tasks in the process of getting relevant technologies to farmers (Swanson, 1997), These are also used:

- To network all concerned and bring them together to share ideas materials, information, expertise by the development chain.
- To institutionalize the mechanisms of information flow, material flow and providing necessary services.
- To ensure continuity of arrangement in the absence of persons.
- To maintain continuous flow of updated/listed information to the client system.
- To meet the unforeseen circumstances and contingencies like floods, cyclones and other natural disasters (drought, pest outbreaks).

The true test of linkage is the achievement of the ultimate goal of increasing production, productivity and profitability within the overall context of the national economy.

## Linkages can be classified into the following types

***Production Linkages :*** to maintain regular flow of information from its source to the client system through govt. and non-govt. agencies including input supply and services.

***Technology Generation Linkages :*** to develop new technologies to meet the emerging problems faced by the client system as well as contingency measures to manage pest outbreaks and natural calamities.

***Post Production Linkages :*** with marketing agencies both for domestic and export market to dove tail the product and services to suit the consumer preferences. Linkages with storage transportation, packaging, agro-industries for value addition.

Biotechnology requires intensive post-production linkages for its wider acceptability among the consumers.

To summarize, a large extent consumer acceptance depends on the type of information that is available. Accordingly, information–its usefulness, quality, accuracy, timeliness and credibility play a pivotal role in consumer behavior influencing both advocacy and the ability of consumers to make informed choices in the market place. Hence, it is essential that government and the biotechnology community ensure access to the balanced and credible information to people. The public will reap the benefits of biotechnology only by weighing the information and deciding that these products contribute to their well-being.

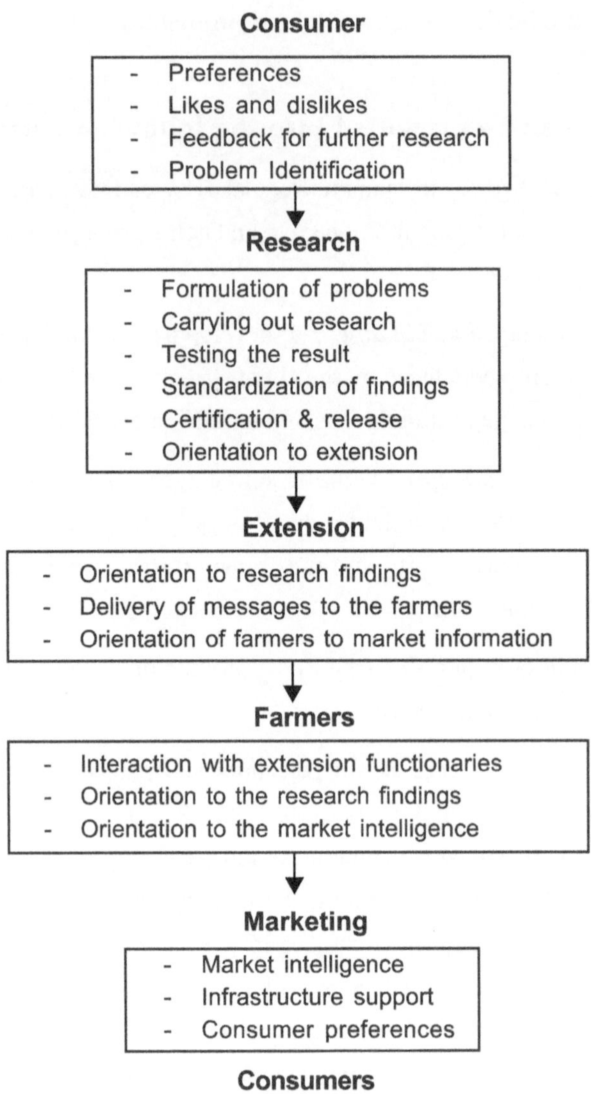


**Fig. 20.1** Consumer-Research-Extension-Farmer-Marketing Linkages

*References*

Bunders Joske, Haverkort Bertus and Hiemstra Wim (editors), Biotechnology – Building on farmers' knowledge, Macmillan Education Ltd, London and Basingstoke, 1996.

Food and Agriculture Organization of the United Nations, Agricultural Biotechnology in the developing world, FAO research and Technology Paper, 6, Daya Publishing House, New Delhi, 1998.

International Center for Agricultural Science and Technology (ICAST), Consumer Acceptance of potatoes that have been genetically modified through Biotechnology, 1994.

Traynor Patrica, Building Public acceptance of agri-food Bioproducts; a comparative international analysis, Virginia Tech, USA, http:// **www.isb.vt.edu/**

# Views of Stakeholders About Agricultural Biotechnology*

**Shri K. Vidyadher Rao**

*Hon'ble Minister for Major Industries, Commerce and Export Promotion, Government of A.P.*

Biotechnology has the potential to provide a wide range of benefits to humanity. It encompasses the whole gamut of human life, starting from Agriculture. The subject influences other aspects like energy, health care, animal husbandary, environmental protection, education and a host of other activities. It is, therefore, imperative that the subject of biotechnology deserves keen attention.

After years of painstaking research and development in biotechnology, the commercialization of this frontier technology for the betterment of mankind is poised to grow exponentially. With the help of biotechnology, new products and processes can be introduced to win a competitive edge over the traditional ones in terms of effectiveness of productivity, cost and safety.

The Government of Andhra Pradesh has declared the area of approximately 600 sq. km. covering 4 mandals in Ranga Reddy District as the "Genome Valley" in which biotech activities will be encouraged and promoted.

This Genome Valley will foster Biotech activities that promote economic development through the growth and exchange of knowledge and technologies. Similar corridors and biotech parks with focus on agriculture, horticulture, marine resources, animal husbandry, forestry and environment protection will be set up in other parts of the state.

* Presented in the workshops on biotechnology for sustainable agriculture organised for different stakeholders.

**Dr. Ummareddy Venkateswarlu**

*Member of Parliament (Lok Sabha) and Chairman, Estimates Committee.*

Green revolution made India self sufficient in food grains and at present 46 million tonnes of buffer stock is stored in the godowns. But we could not be able to export our produce to world market, since the focus of world market is on quality and phyto-sanitary issues. Indian agriculture requires cost effective technologies rather than low cost technologies. Green revolution concentrated in irrigated land, but the effects have not percolated to dryland areas. Agricultural Biotechnology has lot of potential to redress these issues.

Biotechnology with its environmental friendly focus will help to meet the present problems and thrusts. What is needed is institutional framework to channelize the efforts of agricultural scientists for dissemination of biotechnologies, as it provides answer to prevailing evils in the existing agricultural scenario.

The scientific progress in the field is spectacular and it has vast potential for development of sustainable agriculture and evergreen revolution even in drylands. However, some of these technologies have potential benefits as well as risks.

There is need for educating scientists and entrepreneurs about polices and procedures relating to intellectual property rights of biotechnology interventions. In addition, legal experts should acquaint themselves with basics of biotechnology as well as intellectual property rights so that they can interpret cases of patent right infringement in the field. I understand that there is dearth of qualified patent attorneys in India. With implementation of WTO/GATT regulations, more attention needs to be given to this field of study.

**Shri Nimmala Krishtappa**

*Hon'ble Minister for Animal Husbandry and Dairy Development, Government of A.P.*

Biotechnology, has the potential to be a real boon to a country like India, where agriculture is the major profession of the people and where there is a reduction in available land and an increasing demand for food and fibre as a result of increasing population.

The Biotechnology Policy 2001-2002 has identified several objectives and thrust areas of which Agriculture Biotechnology will receive prime attention. The proposed Genome Valley near Hyderabad and Biotech Parks with focus on agriculture, horticulture, animal husbandry. Forestry and environment protection in other parts of the state will help in accelerating the pace of development in agriculture.

In animal husbandry, artificial insemination, development of vaccines for important diseases, diagnostics, cloning of animals, embryo transfer technology are already in practice. However, experimentation is in progress for development of transgenic animals. In order to translate the vision into action, there is need for: (a) accelerating the pace of research in biotechnology, and (b) transferring the technology to the ultimate users. This need revitalization of research, extension and client system. It is widely recognized that awareness about the strengths, weaknesses and potentialities of these technologies is inadequate at present.

**Justice Vaman Rao**

*Retired Judge, AP High Court, Hyderabad*

Under TRIPS agreement, a member country is required to provide for protection of plant varieties either by patents or by an effective *suigeneris* system or any combination thereof. In conformity with this requirement, Indian Parliament has passed. "The Protection of Plant Varieties and Farmers Rights Bill".

This bill seeks to bring about a balance between the rights of breeders, farmers and research workers, which are in some way mutually contradictory.

An in-built provision in the Act, either as a condition of registration or independent provisions mandating the breeder to compensate the farmer at least to some extent appears necessary for a meaningful success of any biotechnology program for improving agriculture.

The same holds good for making a provision to compensate consumers for any consequential health hazards and also for compensation for defiling environment.

Though farmers and consumers do have a right to seek their remedies for compensation under general law, driving them to courts to seek compensation would knock at the bottom of any framework installed for protection of their rights and would prove to be a disincentive for wide acceptance of biotechnologies.

### Justice Bilal Nazki

*Hon'ble Judge A.P. High Court*

The success of biotechnology will depend on its usefulness in improving the living condition of poor people. What good is biotechnology for India if it does not ensure food security, safe drinking water and sanitation facilities for poor masses of India?

There is need for reorienting the priorities of our scientific and educational institutions so that the life of poor people can be improved. The policymakers have to devise policies and programmes for using biotechnological tools for poverty alleviation and environmental management.

### Dr. S.V. Prabhath

*Principal Director, NISIET, Hyderabad.*

There are many causes of hunger, and biotechnology alone cannot solve this problem. But biotechnology is a valuable tool, when used in coordination with efforts to improve food distribution system and diversify diets. It can have a major impact on hunger and malnutrition.

He mentioned, Frary and Tanksley saying "The issue is not whether we should modify the genetics of crop plant. We embarked on that road thousands of years ago when plants were first domesticated. Instead of simply judging the vehicle through which we make genetic changes, we need to weigh the potential consequences that such modifications hold for the society and the environment."

It was unclear how agricultural biotechnology research could better interface with the agricultural extension systems. It cannot be assumed that even agricultural biotechnologies which are wholly and unrestrictedly in the public domain will actually reach the fields of poor farmers in the short term through existing state extension channels.

## Dr. M. Harishchandra Prasad

*Chairman and Managing Director, Andhra Sugars Ltd.*

New field of genetic engineering has no roof, sky is the limit to what benefit mankind could derive with the intelligent use of this technology. Biotechnology should be integrated with traditional technology. The scientists and policy makers should thwart the attempts of multinational companies in introducing terminator gene technology.

The derivatives of Biotechnology should be thoroughly evaluated for environmental impact and to ensure that the GM crops are environmentally as safe as conventional crops. The use of genetically improved crops shall bring down the use of pesticides further and help to preserve the environment.

Scientists should be cautious and adopt stringent procedures to thwart the ill effects of Biotechnology if any, taking into consideration of past five decades experience of indiscriminate use of chemical fertilizers and pesticides which caused all-round pollution and are threatening the existence of the biosphere which has outweighed the benefits derived in using these products.

## Dr. I.V. Subba Rao

*Vice Chancellor, ANGRAU*

India needs to double the food production in next 20 years. It is very difficult to achieve this goal with the existing agricultural technologies. It could be possible, through biotechnology research and development including GMOs.

He made a mention of Dr. Watson's quote, " Never put off doing some thing useful for fear of evil that may never come". Caution need not or should not take the form of prevention.

## Mr. Chengal Reddy

*President, Federation of Farmers Association*

Farmers in countries like China have benefited immensely by using this technology. Indian farmers have had no access to this technology so far. There has been inordinate delay in releasing Bt cotton seeds for cultivation by farmers. In this process, the Indian farming community has been the main loser. Procedures used by Govt. agencies are quite slow and inefficient. There is a need for better coordination among scientific community, private sector companies and farmers for efficient technology transfer from research laboratories to farmer's fields.

# Workshop Recommendations

## I. Public Awareness on Biotechnology for Sustainable Agriculture

1. The workshop resolved that there is need for creating pubic awareness among all stakeholders about agricultural biotechnologies on a sustainable basis through different methods and media. The stakeholders include policy makers, scientists, extension service providers, farmers, consumers and school children.
2. There is dearth of empirical data on cost-benefit ratios of different forms of agricultural biotechnologies that can be used for creating public awareness. Urgent steps are needed to generate this information for a range of field situations.
3. Systematic investigations are needed to understand the current awareness levels, attitudes and perception of farmers towards different forms of agricultural biotechnologies. This information is needed for designing calibrated public awareness for the farmers groups.

4. Institutions dealing with biological, social, economic, managerial and policy issues of agricultural and veterinary biotechnology should join hands in establishing a consortium for the purpose of research, development, extension and evaluation in the field of biotechnology.

5. Awareness building should be done on a continuous basis through collaboration between different institutions. This needs promotion and strengthening of partnership linkages among policy makers, research scientists, extension services providers, farmers, marketing agencies and consumers.

6. The workshop identified the need for serious attempts by experts to demystify biotechnology and develop popular science communication tools so that different groups of stakeholders can participate in the biotechnology debate and dialogue in a realistic way and can make rational decisions.

7. It was recommended that the Biotech Parks should be established at district level so that they can serve as research-cum-training centers for educating farmers. In addition, the opinion leaders, farmer groups and Self Help Groups (SHGs) in the villages could be entrusted with responsibility of maintaining biodiversity registers.

8. The participants stressed the need for formulating a detailed action plan for creating public awareness in A.P. to start with. This should include objectives, activities, members, target groups, institutions/organizations responsible, methods and media, time frame, resources, expected outcome and monitorable indicators. The action plan should contain the partnership frameworks between key institutes/actors based on shared objectives and comparative advantage in support of the state's awareness campaign strategy.

9. The workshop recommended to constitute a core group comprising of representatives from BTU, PRDIS, ANGRAU, ICRISAT, MANAGE, NIRD, NGOs, ICAR institutes, Osmania University, University of Hyderabad, State Departments of Agriculture, Horticulture and Animal Husbandry and farmer representatives to formulate the detailed action plan with commitment of resource to launch the public awareness movement in A.P.

10. There is also need felt by the workshop for identifying one or two nodal agencies (probably one in government sector and another in NGO sector) for promoting public awareness on agriculture biotechnology in a coordinated manner. As an interim arrangement, it was resolved to entrust the responsibility to BTU and PRDIS for taking lead in this direction.

11. In addition to the general recommendations, the workshop deliberated in groups and came out with specific recommendations on broad message subjects, communication strategies (methods and media) to be used in public awareness campaigns for different categories of stakeholders.

## II. Biotechnology for Sustainable Agriculture-Opportunities and Constraints

### A. Public and Private Sector Collaboration

1. There is need for identifying and agreeing on partnership framework between private and public sectors concerning technology generation, building databases and commercializing of products after arriving at the totality of the resources needed.

2. Collaboration between private and public sectors is more specifically to facilitate data sharing on toxicology and shelf-life so that the products can reach the markets directly.

3. There is need for consolidation of research to greater extent towards efficiency, safety and equity.

4. The following areas were identified for attention in public sector institutions for free flow of genetic material to the researchers:

   - Development of parental materials for seed development
   - Transgenic seeds
   - Conversion of materials provided by private sector for use in agricultural and rural development programmes.
   - Outright procurement of technology from abroad on time payment basis (without royalty) by private or public sectors or jointly.

5. Joint patenting (through a memorandum of understanding) can be done by private and public sector institutions in management of IPR during collaborative R&D.

6. Collaborative research between public and private sector institutions may face the constraints in matters pertaining to readily exploitable technologies, training, willingness to share information etc. Hence, there is need to find out solutions through mutual trust and confidence.

7. Development of a policy for sharing information for collaboration and partnership between public and private sector is important for sustainability. There is need for developing:

   - A framework of norms to guide collaborative work
   - Mechanisms for credit sharing between pubic and private sectors

8. NGO's can play an important role in creating public awareness on biotechnologies by implementing bio-village concept, identifying local priorities and communicating them to research institutions and dispelling the apprehension among the people about biotechnology.

## B. Pre and Post release monitoring of GMOs

1. There is need to have a strong network among public and private sector organizations for evaluating and developing biosafety measures before release of GMOs.
2. The areas for introduction of transgenic crops should be identified by Government organizations with the help of farmers organizations.
3. The field trials should be organized simultaneously in different agro-ecological zones instead of conducting them in a phased manner.
4. The long term experiments on the toxicological aspects of GMOs are needed.
5. NGOs should be involved in developing baseline data and monitoring impact of different technologies.
6. Suitable legal framework to protect interest of farmers against risk of GMOs is needed.

## C. Research-Extension-Farmers-Marketing and Consumer Linkages

1. It is recommended to strengthen the linkages by institutionalizing already existing mechanisms through joint planning and implementation plans so that the whole process can operate in partnership mode. While doing so, it is desirable to establish linkage committees (zone wise) and research-extension liaison units (with special emphasis on biotechnology).

2. Ensuring horizontal linkages among interdisciplinary teams of researchers through regular interface mechanism such as training, meetings and outreach programmes is also recommended.

3. It is further observed that utilizing existing windows of opportunities, technology verification, participation in technology scrutiny, marketing and promotion of agri-business are all vital activities that can strengthen linkages and lead to sustainable development.

## D. Biotechnology Products

### 1. Biofertilisers

Three specific areas in respect of Rhizobium were identified for further work.

- Addressing the inconsistencies in results
- Improving competitiveness
- Increasing shelf life
- Information on effectiveness of green manuring using crops like sunhemp, dhaincha etc. need to be widely disseminated.
- There is need for reviewing, compiling and synthesizing information from different sources. The cost-benefit aspects and extent of substitutability of chemical fertilizers by vermi compost should be investigated and the information needs to be widely disseminated.

### 2. Tissue-Cultured Plants

The experience in case of banana shows that cost of tissue-cultured plants restricts adoption among farmers and also the variability in the plant growth is a problem. However, in some cases, the banana farmers have made substantial profit. The experience in case of tissue-cultured teak is mostly positive. The merits and demerits of tissue-cultured plants should be widely disseminated among farmers.

**3. Genetically Modified Crops**

(a) There is need for creating awareness on benefits and possible risks.

(b) The staff in Agriculture department, Agriculture University, NGO sector etc., should be trained about the new technologies so that they can interact with farmers with credible information.

(c) There is need for constituting a core group for compiling and disseminating to farmers with credible information through print as well as electronic media.

**4. General**

1. The clonal propagation with tissue-cultured plants should be encouraged
2. There is need for a certifying agency for proper quality control of all biotechnological products
3. Cost benefit data for various technologies should be generated and shared with farmers.

## III. Biotechnology and Communication Strategies for Sustainable Agriculture

1. The workshop recommended that a media resource and communication centre should be established at state level for creating public awareness on agricultural biotechnology among different stakeholders. It is desirable to start at centre and state level with facilities for production and communication of information materials, aids and other software exclusively for agricultural biotechnologies with the support of BTU and other

partners. Eventually, regional and district centers to address location specific needs could be established. The Agricultural cooperative societies, agro-input dealers, and agri-clinics should be involved in creating public awareness of biotechnologies.

2. The biological control laboratories should be established in each mandal. These laboratories should have linkage at the village level.
3. It was recommended that para-technical workers should be trained in each village for dealing with various biotechnologies.
4. The biotechnology exhibits should be displayed in rythu bazars and farmers' fairs so that the information could be easily disseminated among a large number of farmers.
5. The workshop recommended that field demonstrations should be set up in the farmer's fields at strategic locations. The farmers and extension workers can learn the benefits and limitations of various biotechnology practices through these demonstrations.
6. The farmers' field school (FFS) approach being used for popularizing IPM should be tried for popularization of various biotechnology products and practices on a pilot basis.
7. Private and public extension systems should be equally responsible for biotechnology extension.
8. It is important to form cooperative extension networks at district level with the farmers training centers, KVKs, Government extension service providers, professional NGOs, DAATT centres. ATMA etc to prepare a convergence action plan for public awareness on biotechnologies and also to implement on the ground by rationalizing and harmonizing the resources.

9. The workshop recommended to utilize regional FM broadcasting centers to broadcast location specific information and cases on biotechnologies.

10. The workshop discussed the various prototype audio visual aids and communication media to create public awareness among the extension workers and farmers with appropriate media mix based on the audience preferences and profile.

    (a) Posters

    (b) Charts

    (c) Flash cards

    (d) Flip Boards / Charts

    (e) Leaflets

    (f) Ready Reckoner (questions and answers)

    (g) Radio bulletin

    (h) Slogans and quotations through wall painting, television and radio

    (i) Success stories through publications and mass media

    (j) Video films on each biotechnology

    (k) Folk media

11. It was recommended that the extension service providers should communicate through various methods and media the message concerning not only productivity but also the cost benefits, improvement of quality, marketing aspects vis-a-vis biotechnology application in view of the concerns of WTO / GATT.

12. In addition to the above, the workshop recommended to use the existing uptake pathways such as Rythumitra and Annadata programmes of television, farm radio programms, farmers clubs, village volunteer leaders, self help groups, water user associations, watershed management committees, panchayats etc., to create public awareness on biotechnologies.

## IV. Followup activities at Grass root level:

To meet the challenges, opportunities and problems for followup on biotechnologies at farmers end, certain activities were identified and recommended by the farmer groups.

1. ***Institutionalization*** of biotech activities in the government agencies for follow up without any time lag is needed.
2. ***Biotechnology Unit :*** Establishment of different biotechnology units at mandal/block level for making availability of these technologies.
3. ***Village Adoption :*** Adoption of one or two villages in a mandal / block for intensive research and development adoption and popularization of technologies.
4. ***Training Programmes :*** Organizing training programmes for clonal propagation of tissue culture plants at farmers level.
5. ***Bio-centres :*** Establishment of bio-centres at various levels for providing service to the farmers.
6. **Demonstration:** Organizing demonstrations / farmer field schools (FFS) on biotechnology products and technologies in farmers fields for R & D and to develop confidence among the farmers about these technologies.

7. ***Use of AMCs :*** Utilization of meetings organized at agricultural market committees (AMCs) for creating awareness of various biotechnologies.

8. ***Workshops :*** Organization of workshops on market opportunities of biotech products.

9. ***Providing support price :*** Encouragement of the farmers to adopt these technologies by giving support price for biotech products.

10. ***Production of success stories :*** Locating successful farmers practicing these technologies and producing documentation of success stories for motivating the farmers about these technologies.

11. ***Traditional media :*** Organization of kala jathas and cultural troups in villages for popularizing the technologies and also production of audio cassettes on songs of biotechnology and burrakatha, etc. Projecting traditional media through electronic media for wider public awareness.

12. ***Broadcasting / Telecasting :*** Slogans on biotechnology and benefits of biotech products should be broadcasted / telecasted in between the popular serials. In addition flashing of these items during different programs is recommended.

13. ***Subsidy on seed worms :*** Supply of seed worms on subsidy to encourage the farmers for producing vermicompost.

14. ***Public awareness campaigns :*** Conducting public awareness campaigns by using appropriate visual aids.

## V. Biotechnology for Sustainable Agriculture – The Way Forward

1. There should be more opportunities for exchange of views between policy makers and technical experts. It was emphasized that this interaction process will facilitate optimum utilization of emerging biotechnologies for benefit of the society.
2. The public sector research and technology transfer capacity in the field of agriculture biotechnology should be strengthened. Particular attention should be given to the crops like sorghum, millet, castor etc, which are grown by small farmers in the dry land areas.
3. The Government should strengthen seed supply mechanism for transgenic crops in the rural areas. Special precautions are needed to ensure that even illiterate farmers can get good quality and genuine transgenic seeds.
4. The policy makers should facilitate collaboration between public sector agricultural institutions and private biotechnology companies for achieving a fast pace of research and development for the benefit of farming community.
5. The biotechnology research and development should focus primarily on solving the problems of the poor people.
6. There is need for evolving coordination mechanisms among scientific communities, private sector companies and farmers for efficient transfer of technology from research laboratories to farmers' fields.

# Index

## F



## G



## H



## I



## K



## L

## M



## N



## O



## P



## R



## S



## T



## U



## V

9 789352 300051

Printed at Repro India Ltd.